ENDOCRINE CAUSES OF SEASONAL AND LACTATIONAL ANESTRUS
IN FARM ANIMALS

CURRENT TOPICS IN VETERINARY MEDICINE AND ANIMAL SCIENCE

Control of Reproduction in the Cow, edited by J.M. Sreenan

Patterns of Growth and Development in Cattle, edited by H. de Boer and J. Martin

Respiratory Diseases in Cattle, edited by W.B. Martin

Calving Problems and Early Viability of the Calf, edited by B. Hoffmann, I.L. Mason and J. Schmidt

The Future of Beef Production in the European Community, edited by J.C. Bowman and P. Susmel

Diseases of Cattle in the Tropics: Economic and Zoonotic Relevance, edited by M. Ristic and I. McIntyre

Control of Reproductive Functions in Domestic Animals, edited by W. Jöchle and D.R. Lamond

The Laying Hen and its Environment, edited by R. Moss

Epidemiology and Control of Nematodiasis in Cattle, edited by P. Nansen, R.J. Jørgensen and E.J.L. Soulsby

The Problem of Dark-Cutting in Beef, edited by D.E. Hood and P.V. Tarrant

The Welfare of Pigs, edited by W. Sybesma

The Mucosal Immune System, edited by F.J. Bourne

Laboratory Diagnosis in Neonatal Calf and Pig Diarrhoea, edited by P.W. de Leeuw and P.A.M. Guinée

Advances in the Control of Theileriosis, edited by A.D. Irvin, M.P. Cunningham and A.S. Young

Fourth International Symposium on Bovine Leukosis, edited by O.C. Straub

Muscle Hypertrophy of Genetic Origin and its Use to Improve Beef Production, edited by J.W.B. King and F. Ménissier

Aujeszky's Disease, edited by G. Wittman and S.A. Hall

Transport of Animals Intended for Breeding, Production and Slaughter, edited by R. Moss

Welfare and Husbandry of Calves, edited by J.P. Signoret

Factors Influencing Fertility in the Postpartum Cow, edited by H. Karg and E. Schallenberger

Beef Production from Different Dairy Breeds and Dairy Beef Crosses, edited by G.J. More O'Ferrall

The Elisa: Enzyme-Linked Immunosorbent Assay in Veterinary Research and Diagnosis, edited by R.C. Wardley and J.R. Crowther

Indicators Relevant to Farm Animal Welfare, edited by D. Smidt

Farm Animal Housing and Welfare, edited by S.H. Baxter, M.R. Baxter and J.A.D. MacCormack

Stunning of Animals for Slaughter, edited by G. Eikelenboom

Manipulation of Growth in Farm Animals, edited by J.F. Roche and D. O'Callaghan

Latent Herpes Virus Infections in Veterinary Medicine, edited by G. Wittmann, R.M. Gaskell and H.-J. Rziha

Grassland Beef Production, edited by W. Holmes

Recent Advances in Virus Diagnosis, edited by M.S. McNulty and J.B. McFerran

The Male in Farm Animal Reproduction, edited by M. Courot

Endocrine Causes of Seasonal and Lactational Anestrus in Farm Animals, edited by F. Ellendorff and F. Elsaesser

ENDOCRINE CAUSES OF SEASONAL AND LACTATIONAL ANESTRUS IN FARM ANIMALS

A Seminar in the CEC Programme of Co-ordination of Research
on Livestock Productivity and Management, held at the Institut
für Tierzücht und Tierverhalten, Mariensee,
Bundesforschungsanstalt für Landwirtschaft (FAL),
October 2–3, 1984

Sponsored by the Commission of the European Communities,
Directorate-General for Agriculture, Co-ordination of Agricultural
Research

Edited by

F. Ellendorff, F. Elsaesser
Institut für Tierzücht und Tierverhalten, Mariensee
Federal Republic of Germany

1985 **MARTINUS NIJHOFF PUBLISHERS**
a member of the KLUWER ACADEMIC PUBLISHERS GROUP
DORDRECHT / BOSTON / LANCASTER
for
THE COMMISSION OF THE EUROPEAN COMMUNITIES

Distributors

for the United States and Canada: Kluwer Academic Publishers, 190 Old Derby Street, Hingham, MA 02043, USA

for the UK and Ireland: Kluwer Academic Publishers, MTP Press Limited, Falcon House, Queen Square, Lancaster LA1 1RN, UK

for all other countries: Kluwer Academic Publishers Group, Distribution Center, P.O. Box 322, 3300 AH Dordrecht, The Netherlands

Library of Congress Cataloging in Publication Data

```
Main entry under title:

Endocrine causes of seasonal and lactational anestrus
    in farm animals.

    (Current topics in veterinary medicine and animal
science)
    1. Anestrus--Congresses.  2. Veterinary
endocrinology--Congresses.  I. Ellendorff, F.
II. Elsaesser, F.  III. C.E.C. Programme of Coordination
of Research on Livestock Productivity and Management.
IV. Commission of the European Communities.  Coordination
of Agricultural Research.  V. Series.
SF768.3.E54  1985        636.089'8172        85-10646
```

ISBN-13: 978-94-010-8726-1   e-ISBN-13: 978-94-009-5026-9
DOI: 10.1007/978-94-009-5026-9

EUR 9554 EN

Book information

Publication arranged by: Commission of the European Communities, Directorate-General Information Market and Innovation, Luxembourg

Copyright/legal notice

CONTENTS

Page

P R E F A C E

The EC-Seminar "Endocrine Causes of Seasonal and Lactational Anestrus in Farm Animals" took place October 2 and 3, 1984 at the Institut für Tierzucht und Tierverhalten, Mariensee, Bundesforschungsanstalt für Landwirtschaft (FAL).

Season as well as parturition and lactation result in absence or reduction of fertility leading to a considerable loss of revenue in animal production. Correction or prevention of such naturally occurring states of infertility would markedly increase efficiency and potential of reproduction in all major livestock species. Although severe physiological barriers have been present, there are increasing activities within common market countries to investigate causes and possible corrective measures to overcome such periods of reduced fertility.

This seminar was designed to pursue several objectives:

(1) To discuss and summarize present knowledge of causes and techniques related to seasonal and lactational anestrus;

(2) to bring EC-laboratories working in the field closer and to stimulate further coordination or collaboration;

(3) to stimulate new research in this area.

To this end the program headings were selected as (a) Seasonal Anestrus; (b) Lactational Anestrus; (c) Post partum Anestrus. The species discussed were sheep, cow, horse and pig. Basic and applied endocrine aspects of seasonal and lactational anestrus were considered in each session.

The seminar has to a large extent fulfilled its objectives and it has contributed considerably to our understanding of naturally occurring periods of reduced fertility in our farm animals. It has also shown directions that may eventually lead to its control and thus increase efficiency of animal production.

The EC gratefully acknowledges the contributions of the participants and the efforts of the organizers. The seminar helped to further promote interdisciplinary research in the area of animal reproduction and the cooperation within the European Communities.

SESSION I

SEASONAL ANESTRUS IN THE EWE

Chairman: E. Lamming

SEASONAL ANESTRUS - PROBLEMS AND PERSPECTIVES

G.E. Lamming

AFRC Research Group on Hormones and Farm Animal Reproduction,
Faculty of Agricultural Science, University of Nottingham,
Sutton Bonington, Loughborough, Leics., U.K.

The Problem

It can be assumed that the most appropriate reproductive strategy for
any mammalian species is that sequence of events involving ovulation,
mating, pregnancy and lactation most likely to favour the long-term
survival of species. For those smaller mammals, where gestation and
lactation lengths are short, the individual female is able to produce
more than one offspring or litter per annum. However, for several
obvious reasons man has utilized mainly the larger herbivores as source
of animal protein and as body size increases, gestation lengths are
extended. It then becomes impossible to accommodate successive pregnancies
within one annual cycle. In this situation lactation is usually followed
by a period of seasonal anovulation. Indeed the wild ungulate ancestors
from which our domestic species have evolved are basically seasonal
breeders. They have developed a response pattern to strong environmental
cues to control estrus and ovulation. It also follows that the responses
to these cues to show estrus and ovulation are more forceful and distinct
if they are contrasted with the absence of behavioural estrus and ovulation at other periods.

For at least a millenium man has attempted to manipulate the reproductive patterns of his mammals in order to remove these constraints. Our
problem is that he has been largely unsuccessful. He has resorted to the
application of environmental cues (manipulation of light patterns) or
hormone treatments to manipulate their reproductive pattern. The problem
of control has been ex acerbated since we are also forcing animals to
perform at higher latitudes from their origins, where, due to climatic
factors and the seasonal variations in the availability of food they
experience extended anovulatory periods. Thus I believe, the successful
strategy of the ancestors of our domesticated ungulates probably involved
the establishment of a market contrast between anovulation and anestrus
during the non-breeding season and the free running ovarian cycles during

4

the breeding season, thus adjusting the time of ovulation so that parturition occurs at the most appropriate time for survival.

The History of Manipulating the Cycle with Hormones

Since the 1930's when gonadotrophins were discovered and the potential use of PMSG was established we have seen a number of attempts to manipulate the cycle in cattle, sheep and pigs by the use of PMSG-steroid combinations. The investigations were expanded when intravaginal application of steroid hormones became possible. We had to rely on PMSG because the gonadotrophins have not been available in sufficient quantities or at a sufficient purity to permit their widescale use. Furthermore, although the French and Irish workers have achieved success in sheep with the use of PMSG-progesterone sponge treatments a judicious choice of breed is essential for success (i.e. those with a long breeding season and high inherent fecundity - the Finn x Dorset Horn, or Romanov crosses). We need a wider range of success. Clearly the steroid hormone feedback control systems influencing hypothalamic and pituitary function operate with a fine degree of control and the empirical administration of the active steroid hormones or synthetic analogues together with PMSG does not always produce a reliable response, either because the declining level of the "steroid" component does not simulate natural luteolysis or there is a variable response to the PMSG-component for reasons not well understood. Such treatments fail to cope with the intricacies of the endogenous control systems in different breeds or in variable environmental situations.

The Prospectives

With increasing use of recombinant DNA technology we can forsee an increased provision of cheap sources of highly purified animal hormones or active hormone fragments. This has already occurred with bovine "growth hormone" fragments. We already have adequate supplies of the naturally occurring steroid hormones but we need more information on how to use them.

However, to increase our control of seasonal anovulation we need more detailed information on the nature of the responses our sheep make to these important environmental and behavioural cues. We need to know more about the sensory signals to the central neuronal control areas and how the neural signals are interpreted and the nature of the effector mechanisms. Only then will we be able to define more precisely:

1) the nature of the photoperiodic response in different species and breeds and the effects of interaction between the sexes;

2) the pattern of hormone changes which distinguish between the breeding and non-breeding season.

The following contribution will address these issues.

ENDOCRINE BASIS OF SEASONAL ANOESTRUS IN SHEEP

W. Haresign, B. J. McLeod, G. M. Webster, K. Worthy

University of Nottingham School of Agriculture,
Sutton Bonington, Loughborough, Leics LE12 5RD, U.K.

ABSTRACT

An increase in the negative feedback effects of oestradiol on tonic LH secretion, coinciding with the period of seasonal anoestrus, is thought to be casually related to the seasonal nature of breeding in the ewe. This results in a low frequency of episodic LH secretion, below that required to promote the final phases of follicle development and maturation which immediately preceed ovulation. Artificially increasing tonic LH secretion in seasonally anoestrous ewes, by administration of low doses of GnRH in the form of either pulsed injections or continuous infusion, will induce follicle development with an associated increase in oestradiol secretion sufficient to promote the positive feedback response and thereby induce ovulation. This occurs without the need for any GnRH-induced increase in FSH secretion. A high proportion of the induced corpora lutea are functionally abnormal unless the GnRH treatment is preceeded by a period of progesterone priming.

INTRODUCTION

The seasonal nature of reproduction imposes a major constraint on attempts to improve the reproductive efficiency of sheep flocks. Seasonal breeding may be explained as an evolutionary adaptation to ensure the most beneficial environment for the survival of the new-born lamb in the wild, but modern husbandry systems can overcome these environmental problems such that there is now considerable interest in methods to increase the breeding frequency of sheep flocks above the conventional one lamb crop/annum. The wide range of environments in which sheep have developed has resulted in considerable variation in the length of the breeding season, ranging from the monoestrous condition of some primitive breeds to those (e.g. Merino) which are capable of producing lambs in most months of the year (Eckstein and Zuckerman, 1956). As well as between-breed variation in breeding season length, there is also evidence of considerable variation between ewes within a breed (Hafez, 1952; G. M. Webster and W. Haresign, unpublished data). Furthermore, it has been suggested that the depth of seasonal anoestrus also varies between breeds, and is positively correlated with the length of the non-breeding interval (Robinson, 1951). Unfortunately, little comparative work has been conducted with modern hormone assay systems to determine whether such differences in depth of anoestrus are reflected in quantitative differences in the patterns of gonadotrophin secretion (e.g. frequency of LH episodes).

PATTERNS OF HORMONE SECRETION DURING THE OESTROUS CYCLE OF THE EWE.

A schematic diagram of the changes in hormone concentrations that occur throughout the oestrous cycle is presented in Figure 1. Recent evidence from many laboratories has indicated that, between successive preovulatory gonadotrophin surges, the pattern of LH secretion is episodic in nature. During the luteal phase of the cycle the high progesterone titre, in synergism with low levels of oestradiol, exert feedback on the hypothalamic-pituitary axis to limit LH episode frequency to one pulse/3 to 12h. However, following regression of the corpus luteum, and thus the removal of progesterone from the system, episode frequency gradually

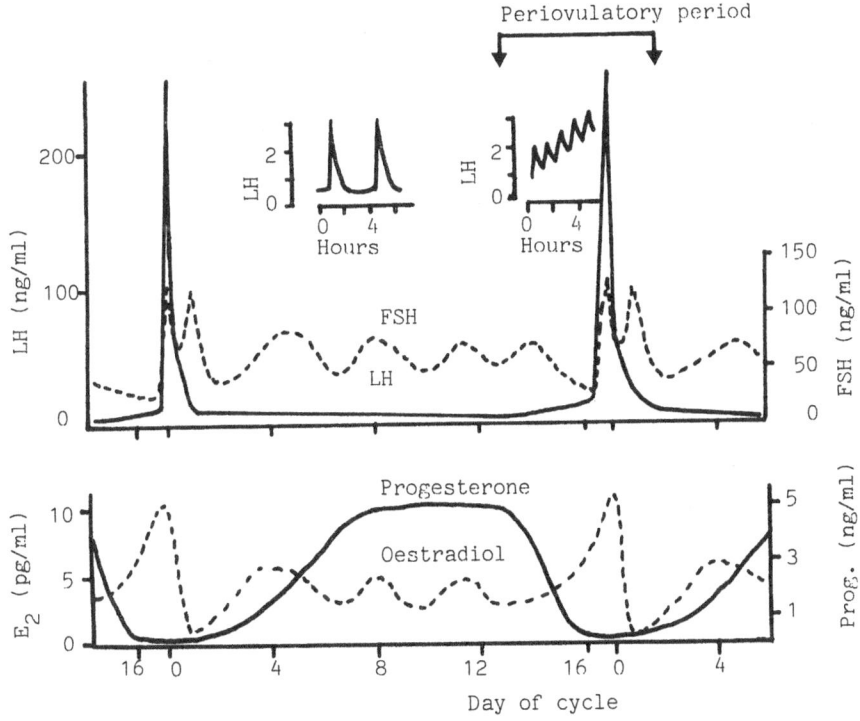

Figure 1.   Schematic diagram of the pattern of change in plasma hormone concentrations throughout the oestrous cycle in the ewe.

increases until it attains a rate of one episode/ 1 to 2h immediately prior to the preovulatory gonadotrophin surge (Baird, Swanston and Scaramuzzi, 1976; Hauger, Karsch and Foster, 1977; Baird, 1978). An inverse relationship between tonic FSH secretion and progesterone concentrations throughout the cycle is far less evident; indeed FSH concentrations show irregular periods of relatively high values interspersed with periods of low levels throughout the luteal phase and then consistently decline during the follicular phase (Cahill, et al., 1981; Lahlou-Kassi, Schams and Glatzel, 1984). This differential pattern of change in tonic LH and FSH concentrations exists because, while tonic LH concentrations can be fully accounted for by the negative feedback effects of ovarian steroids, the control of FSH secretion requires an additional non-steroidal ovarian component, thought to be inhibin (Goodman, Pickover and Karsch, 1981).

The parallel increases in both oestradiol and tonic LH secretion over the follicular phase of the cycle (Baird, 1978; Karsch, et al., 1978) has led to the suggestion that the increase in episode frequency that follows luteal regression is responsible for driving the final phases of follicle growth and maturation. Evidence from other species indicates that FSH, in conjunction with low levels of oestradiol, is required to stimulate the earlier phases of follicle growth and promote the development of LH receptors in both the thecal and granulosa layers, thereby allowing the follicle to respond to the follicular phase increase in tonic LH secretion. Circumstantial evidence supports the existence of a similar mechanism in the sheep (see review by Haresign, 1985).

PATTERNS OF HORMONE SECRETION DURING SEASONAL ANOESTRUS

Throughout seasonal anoestrus there is an absence of preovulatory surges of FSH and LH, and thus of ovulation, with the result that plasma progesterone concentrations remain basal (Roche et al., 1970; Yuthasastrakosol, Palmer and Howland, 1975). This is not due to the lack of a competent positive feedback mechanism, since the balance of evidence suggests that sensitivity of the hypothalamic-pituitary axis to oestradiol is unchanged (Beck and Reeves, 1973; Goodman et al., 1981; Haresign and Friman, 1983). However, comparison of the patterns of tonic LH secretion during seasonal anoestrus and the breeding season shows that the frequency of LH episodes during anoestrus is lower than that recorded during the luteal phase of the cycle (Scaramuzzi and Martenz, 1975; Scaramuzzi and Baird, 1979).

Much less is known of the patterns of FSH secretion during seasonal anoestrus, but circumstantial evidence from several sources suggests that the levels are similar to those recorded during the luteal phase of the oestrous cycle (McNeilly, O'Connell and Baird, 1982; G.M. Webster, B. J. McLeod and W. Haresign, unpublished data). Such data support the suggestion that anoestrus is due to an inadequate pattern of tonic LH secretion for promoting the final phases of follicle growth and development, with the result that the concentrations of oestradiol required to provide the positive feedback trigger are never attained.

Using the ovariectomized, steroid-treated ewe, Karsch et al. (1978) proposed a working hypothesis for seasonal breeding, which has as its basis a seasonal shift in the responsiveness of the hypothalamic-pituitary axis to the negative feedback effects of oestradiol. Following ovariectomy during anoestrus, oestradiol alone is an effective inhibitor of tonic LH secretion, whereas during the breeding season progesterone is required in concert with oestradiol to inhibit the post-castration rise in LH concentrations. Anoestrus will therefore exist for as long as oestradiol alone is able to maintain tonic LH secretion below the threshold required to drive the final phases of follicle growth and development. These observations have recently been extended by work in our own laboratory, designed to determine whether the same relatonships exist for breeds of ewe with widely different breeding season lengths, and whether the seasonal change in prolactin secretion might be involved in mediating this seasonal shift in negative feedback responses (Webster and Haresign, 1983; Worthy and Haresign, 1983 and unpublished data). The data presented in Figure 2 indicate that, even for ewes with widely different breeding season characteristics, there is a marked increase in LH concentrations in ovariectomised, oestrogen-treated ewes, which is coincident with the onset of the breeding season in entire ewes of the same breed. Conversely, entire ewes ceased cyclical activity when LH concentrations in the ovariectomised, oestrogen-treated ewes were approaching basal values. Other studies have indicated that the marked seasonal variation in prolactin concentrations is not a major determinant of this seasonal change in negative feedback responses to oestradiol (Worthy and Haresign, 1983 and unpublished data).

10

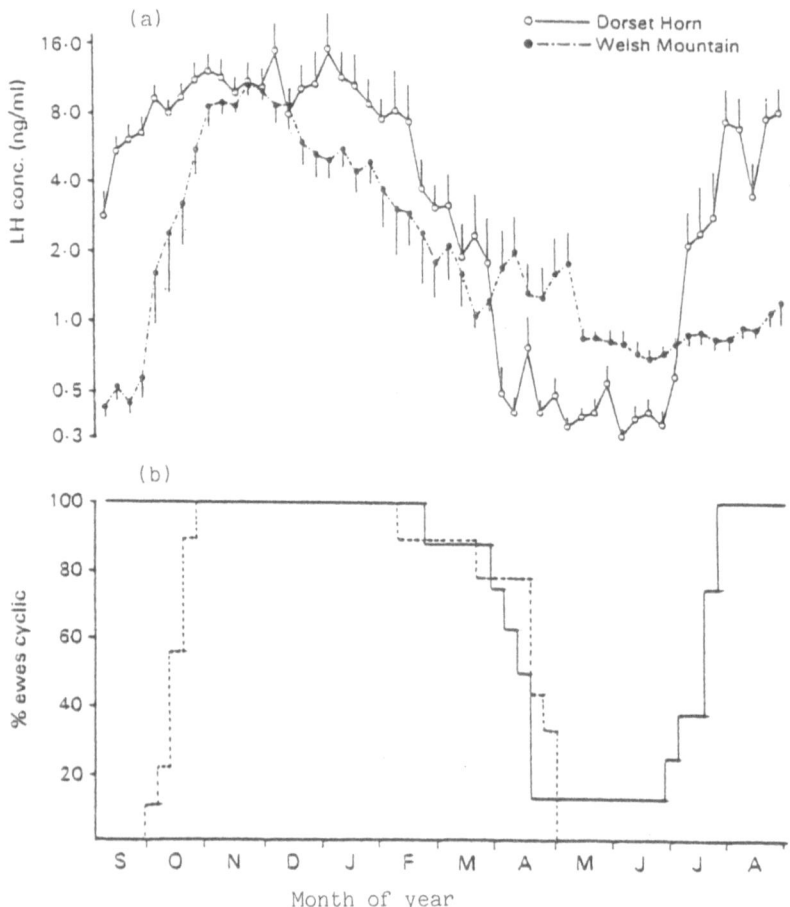

Figure 2. Mean (± s.e.m.) plasma LH concentrations (a) in
          ovariectomized, oestrogen-implants Dorset Horn
          and Welsh Mountain ewes throughout the year,
          and cyclic-activity (b) of untreated entire ewes
          of the same two breeds (after Webster and Haresign,
          1983).

Analysis of the patterns of tonic LH secretion in 24-hour hormone
profiles collected at different times throughout the year has indicated
that the change in mean concentrations depicted in Figure 2 are directly
related to changes in both the frequency and amplitude of LH pulses (G.M.
Webster, K. Worthy and W. Haresign, unpublished data). When LH episodes
occurred at the rate of one/2h in ovariectomised, oestrogen-treated ewes,
a frequency which is capable of promoting follicle development and

ovulation in seasonally anoestrous ewes (see later section), the corresponding weekly mean concentration was 2 ng/ml. If it is assumed that ovariectomised, oestrogen-treated ewes of each breed would show cyclic activity for as long as their weekly mean LH concentrations exceeded 2 ng/ml, an "assumed breeding season length" for individual ewes can be calculated. When this was compared to the actual figures recorded in entire animals, there was no significant difference between the actual and assumed mean length of the breeding season for each breed type, nor in the degree of variation about that mean (G.M. Webster and W. Haresign, unpublished data). Such an observation lends support to the concept that a seasonal shift in negative feedback responses to oestradiol underlies the seasonal pattern of breeding in the ewe, although little is yet known of the precise mechanisms which bring about such a change within the animal. However, the very close association between the pulsatile patterns of GnRH secretion in hypothalamic-hypophyseal portal blood and pulsatile LH secretion in jugular vein blood (Clarke and Cummins, 1982) suggests that anoestrus results from a deficiency at the hypothalamic level.

RESPONSES TO MANIPULATON OF GONADOTROPHIN SECRETION WITH GnRH IN ANOESTROUS EWES.

An extensive series of experiments has therefore been conducted to monitor the gonadotrophin response of seasonally anoestrous ewes to low-dose GnRH therapy, and its effect on ovarian activity.

Pulsed injecton of GnRH

Figure 3 and Table 1 present the results of an experiment in which the LH pulse frequency of seasonally anoestrous ewes was increased from a pre-treatment mean value of 1.5 ± 0.2 episodes/12h up to a value of 1 episode/2h by the administration of GnRH (McLeod, Haresign and Lamming, 1982). A range of doses of GnRH (75-500 ng/injection), designed to produce LH episodes of similar magnitude to those observed during the pretreatment period, were administered intravenously at 2-hourly intervals for 48h. Such treatment eventually culminated in a preovulatory LH surge which was coincident with the occurence of oestrus in progesterone-primed animals. This suggests that the preovulatory surge was due to the

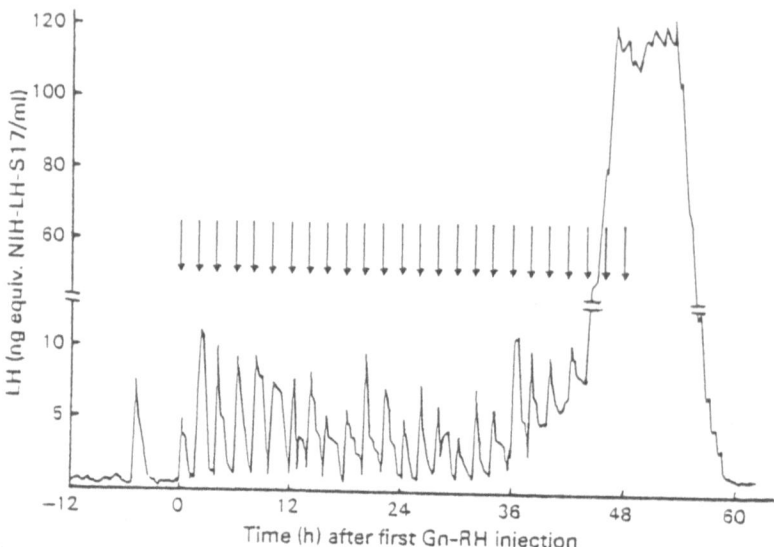

Figure 3. Plasma LH concentrations in a progesterone-primed
seasonally anoestrous ewe treated with 2-hourly
injections (↓) of 250 ng GnRH for 48h (after
McLeod, Haresign and Lamming, 1982).

TABLE 1. Ovulation rate, timing of the preovulatory LH peak and the
incidence of normal luteal function in seasonally aoestrous ewes
induced to ovulate by giving repeated injections of low doses of
GnRH at 2-hourly intervals for 48h, with (Group 5) or without (Groups
1-4) a 14 day period of progesterone priming.

Group	Dose of GnRH (ng/injection)	Ovulation rate	Timing from 1st 1st injection to LH peak (h)	Incidence of normal luteal function
1	75	1.20±0.20	21.80±6.99	1/5
2	125	1.40±0.24	22.90±5.99	2/5
3	250	1.80±0.37	17.70±0.70	1/5
4	500	1.50±0.29	20.40±1.48	1/4
5	250	1.73±0.15	33.89±1.75	15/15

positive feedback effects of oestrogen from developing follicles, rather than being a direct response to the GnRH injections. Analysis of the plasma FSH profiles of GnRH treated ewes indicated that a GnRH-induced increase in FSH concentrations was not required; indeed FSH concentrations over the period from the start of GnRH therapy to the onset of the preovulatory LH and FSH surge declined (Figure 4), in a similar manner to that reported to occur during the follicular phase of the oestrous cycle (see earlier section).

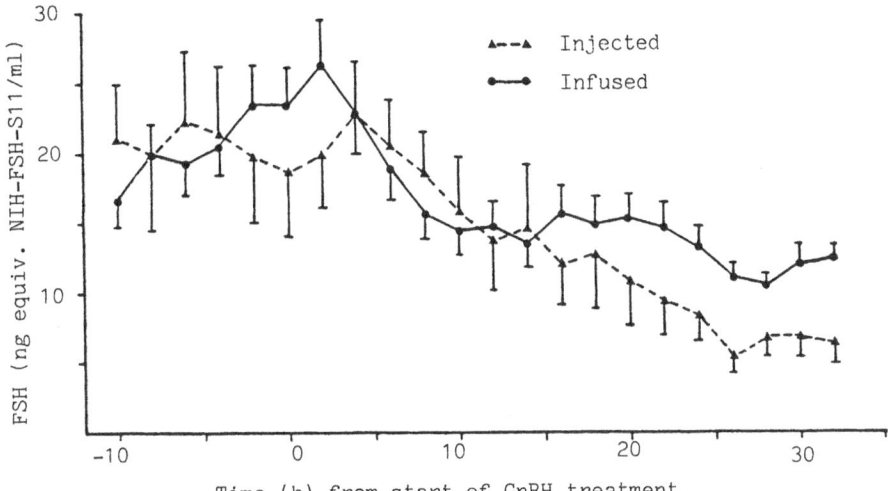

Time (h) from start of GnRH treatment

Figure 4. Mean ($\pm$ s.e.m.) plasma FSH concentrations in seasonally anoestrous ewes injected (N=22) or infused (N=9) with low doses of GnRH for 48h.

Although 34 of the 35 ewes treated with GnRH ovulated, a high proportion of the corpora lutea were functionally abnornmal (Table 1). Only 5/19 ewes induced to ovulate with GnRH alone produced normal corpora lutea, compared with all 15 of the ewes treated with progesterone for 14 days before the start of GnRH injections. In addition, the induced ovulation in ewes pretreated with progesterone was fertile, since all animals came into oestrus, were hand-mated once to entire rams, and 11/15 subsequently lambed at term (McLeod and Haresign, 1984a). When compared over all treatment groups, the preovulatory LH surge occurred significantly (P<0.001) later (35.9 $\pm$ 1.5h) in relation to the start of GnRH treatment in those ewes showing normal luteal function compared to

14

those in which the corpus luteum was defective (22.0 ± 1.1h). Recent data
indicate that the abnormal corpora lutea do secrete small amounts of
progesterone between 2 and 4 days after ovulation and are histologically
indistinguishable from normal at this time, but then undergo rapid
regression (J.A. Southee, M. G. Hunter and W. Haresign unpublished data).

Continuous infusion of GnRH

In an attempt to answer the question of whether it is the increase LH
episode frequency that is the pre-requisite for ovarian follicle
development, or whether it is the effective increase in mean tonic LH
concentrations that such an increase in frequency produces which is
important, progesterone-primed seasonally anoestrous ewes have been
continually infused with either 125 or 250 ng GnRH/h for 48h. Although
such treatment resulted in an immediate elevation in tonic LH
concentrations which was sustained until the onset of the preovulatory LH
surge (Figure 5), the concentrations still exhibited a pulsatile-type
pattern of secretion. Whether such fluctuations are important at the
ovarian level has therefore yet to be determined. Nonetheless,
continuous administration was as effective as pulsatile administration of
GnRH in inducing ovulation and normal luteal function (McLeod, Haresign
and Lamming, 1983), and acceptable conception rates to a single mating
were obtained (McLeod and Haresign, 1984a). Again, GnRH treatment was
associated with a decrease in plasma FSH concentrations from the start of

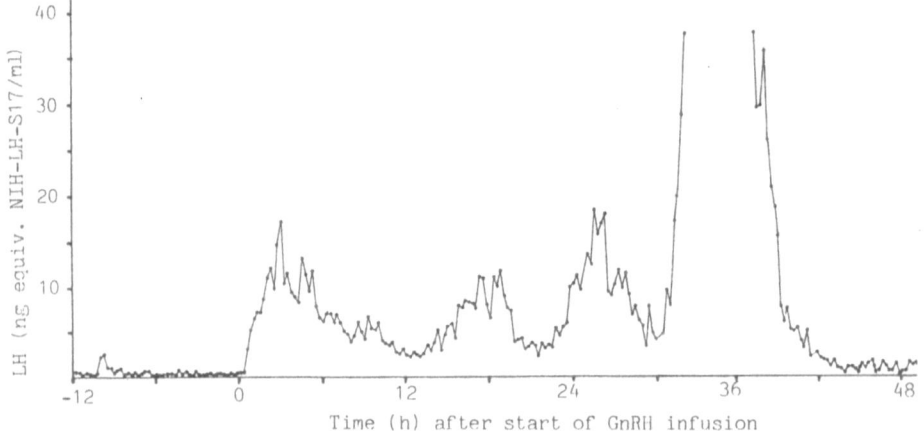

Figure 5.   Plasma LH concentrations in a progesterone-primed seasonally
anoestrous ewe infused with 250 ng GnRH/h for 48h.

infusion until the time of the preovulatory LH and FSH surges (Figure 4; B.J. McLeod and W. Haresign, unpublished data), indicating that the ovarian responses obtained were a function of induced changes in the pattern of LH secretion alone.

Effect of progesterone on preovulatory follicle development in anoestrous ewes

The results presented in Table 1 clearly illustrate a beneficial effect of progesterone-priming on subsequent luteal development, although they provide little information on its possible mode of action. A similar relationship has also been found to exists between progsterone-priming and the incidence of normal luteal function following ram-induced ovulations (Cognie et al., 1982). With both treatments, progesterone-priming is associated with a delay in the timing of the preovulatory LH surge. However, this does not appear to be critical in determining whether luteal function will be normal, since inducing an early LH surge in progesterone-primed anoestrous ewes treated with small doses of GnRH still produces functionally normal corpora lutea in all animals (McLeod and Haresign, 1984b). Furthermore, the fact that only short durations of progesterone-priming are required to ensure normal luteal development in anoestrous ewes induced to ovulate with either pulsatile GnRH therapy (McLeod and Haresign, 1984b) or following the 'ram-effect' (Cognie et al., 1982), and that a single injection of 20 mg progsterone can be given up to 5 days prior to ram exposure and yet still be efficacious (W. Haresign, D.T. Pearce, C. M. Oldham and D.R. Lindsay, unpublished data), all indicate that progsterone may have a direct effect at the ovarian level. This suggestion has received support from recent studies designed to moniter preovulatory follicle development using a variety of in vitro techniques in anoestrous ewes treated with repeated injections of low doses of GnRH. Although there were no differences recorded in any parameter of follicle development during the period from the start of treatment until the onset of the preovulatory LH surge, within 2h of the start of the surge those ewes primed with progesterone produced preovulatory follicles which secreted significantly more oestradiol into the culture medium, and had a significantly greater complement of LH receptor activity in the granulosa layer compared to those treated with GnRH alone (Table 2; M. Hunter, J.A. Southee, B. J.

McLeod and W. Haresign, unpublished data).

TABLE 2. Effect of progesterone-priming on oestradiol secretion rates and granulosa cell LH receptor levels of preovulatory follicles collected around the time of the start of the preovulatory LH surge from seasonally anoestrous ewes treated with pulsed injections of GnRH (M.G. Hunter, J.A. Southee and W. Haresign, unpublished data).

	Progesterone-primed		No progesterone-priming		
	Time of collecton of follicles (h from start of LH surge)				
	0	+2	0	+2	s.e.d.
Oestradiol secretion rate (ng/ml/2h)	3.90	11.93	5.33	3.98	2.07
Granulosa cell I^{125}_3-hCG* binding (cpm x 10^3)	18.1	35.3	13.2	7.5	7.4

* The capacity to bind I^{125}-hCG was taken as an estimate of LH receptor activity.

CONCLUSIONS

The data presented indicate that seasonal anoestrus is a period during which there is a high sensitivity of the hypothalamic-pituitary axis to the negative feedback effects of oestradiol on tonic LH secretion. This results in a reduction in LH episode frequency to below that required to drive the final stages of follicle growth and development. Correcting this inadequacy by appropriate GnRH therapy will promote the full sequence of hormonal changes necessary to induce ovulation, although a period of progesterone priming is required to ensure normal luteal function in all animals.   To date most studies have concentrated on breeds of ewe which have a moderately shallow seasonal anoestrus.   Whether or not such short-term treatments will be as efficacious in deeply anoestrous breeds (eg. Welsh Mountain or Scottish Blackface) requires further study.   It is possible, for example, that in deeply anoestrous breeds tonic FSH is also depressed, such that there are no follicles at an appropriate stage of development to respond to an increase in tonic LH secretion, and that the treatments may need to be modified for such ewes.

REFERENCES

Baird, D.T. 1978. Pulsatile secretion of LH and ovarian estradiol during the follicular phase of the sheep estrous cycle. Biol. Reprod. 18, 359-364.

Baird, D.T., Swanston, J. and Scaramuzzi, R.J. 1976. Pulsatile release of LH and secretion of ovarian steroids in sheep during the luteal phase of the estrous cycle. Endocrinology 98, 1490-1496.

Beck, T.W. and Reeves, J.J. 1973. Serum luteinizing hormone (LH) levels in ewes treated with various dosages of 17β-estradiol at three stages of the anestrous season. J. Anim. Sci. 36, 566-570.

Cahill, L.P., Saumande, J., Ravault, J.P., Blanc, M., Thimonier, J., Mariana, J.C. and Mauleon, P. 1981. Hormonal and follicular relationships in ewes of high and low ovulation rates. J. Reprod. Fert. 62, 141-150.

Clarke, I.J. and Cummins, J.T. 1982. The temporal relationship between gonadotrophin releasing hormone (GnRH) and luteinizing hormone (LH) secretion in ovariectomized ewes. Endocrinology 111, 1737-1739.

Cognie, Y., Gray, S.J., Lindsay, D.R., Oldham, C.M., Peace, D.T. and Signoret, J.P. 1982. A new approach to controlled breeding in sheep using the "ram effect". Proc. Aust. Soc. Anim. Prod. 14, 519-522

Eckstein, P. and Zuckerman, S. 1956. In: Marshalls Physiology of Reproduction, Vol. 1. (Ed. A. S. Parkes). (Longmans, Green and Co., London) pp. 245-249.

Goodman, R.L., Legan, S.J., Ryan, K.D., Foster, D.L. and Karsch, F.J. 1981. Importance of variations in behavioural and feedback actions of oestradiol to the control of seasonal breeding in the ewe. J. Endocr. 89, 229-240.

Goodman, R.L., Pickover, S.M. and Karsch, F.J. 1981. Ovarian feedback control of follicle-stimulating hormone in the ewe: evidence for selective suppression. Endocrinology 108, 772-777.

Hafez, E.S.E. 1952. Studies on the breeding season and reproduction in the ewe. J. agric. Sci., Camb. 42, 189-265.

Haresign, W. 1985. The physiological basis for variation in ovulation rate and litter size in sheep. Livestock Prod. Sci. (in press).

Haresign, W. and Friman, B.R. 1983. Response of ovariectomised ewes to injection of oestradiol - 17β at different times of the year. J. Reprod. Fert. 69, 469-472.

Hauger, R.L., Karsch, F.J. and Foster, D.L. 1977. A new concept for control of the estrous cycle in the ewe based on the temporal relationships between luteinizing hormone, estradiol and progesterone in peripheral serum and evidence that progesterone inhibits tonic LH secretion. Endocrinology 101, 807-817.

Karsch, F.J., Legan, S.J., Ryan, K.D. and Foster, D.L. 1978. The feedback effects of ovarian steroids on gonadotrophin secretion. In: Control of Ovulation. (Eds. D. B. Crighton, N.B. Haynes, G.R. Foxcroft and G. E. Lamming). (Butterworths, London) pp. 29-48.

Lahlou-Kassi, A., Schams, D and Glatzel, P. 1984. Plasma gonadotrophin concentrates during the oestrous cycle and after ovariectomy in two breeds of sheep with low and high fecundity. J. Reprod. Fert. 70, 165-173.

McLeod, B.J. and Haresign, W. 1984a. Induction of fertile oestrus in seasonally anoestrous ewes with low doses of GnRH. Anim. Reprod. Sci. 7, 413-420.

18

McLeod, B.J. and Haresign, W. 1984b. Evidence that progesterone may influence subsequent luteal function by modulating preovulatory follicle development. J. Reprod. Fert. 71, 381-386.

McLeod, B.J., Haresign, W. and Lamming, G.E. 1982. Response of seasonally anoestrous ewes to small-dose multiple injections of Gn-RH with and without progesterone pretreatment. J. Reprod. Fert. 65, 223-230.

McLeod, B.J., Haresign, W. and Lamming, G.E. 1983. Induction of ovulation in seasonally anoestrous ewes by continuous infusion of low doses of GnRH. J. Reprod. Fert. 68, 489-495.

McNeilly, A.S., O'Connell, M. and Baird, D.T. 1982. Induction of ovulation and normal luteal function by pulsed injections of luteinizing hormone in anestrous ewes. Endocrinology, 110, 1292-1299.

Roche, J.F., Foster, D.L., Karsch, F.J., Cook, B. and Djuik, P.J. 1970. Levels of luteinizing hormone in the sera and pituitary of ewes during the estrous cycle and anestrus. Endocrinology 86, 568-572.

Scaramuzzi, R.J. and Baird, D.T. 1979. Ovarian steroid secretion in sheep during anoestrus. In: Sheep Breeding, 2nd Edition. (Ed. G. J. Tomes, D.E. Robertson, R. J. Lightfoot and W. Haresign). (Butterworths, London). pp. 463-470.

Scaramuzzi, R.J. and Martenz, N.D. 1975. Effects of active immunisation against androstenedione on luteinizing hormone levels in the ewe. In: Immunization with Hormones in Reproductive Research. (Ed. E. Nieschlag) (North Holland Publishing Co., Amsterdam) pp.141 - 147.

Webster, G.M. and Haresign, W. 1983. Seasonal changes in LH and prolactin concentrations in ewes of two breeds. J. Reprod. Fert. 67, 465-471.

Worthy, K. and Haresign, W. 1983. Evidence that the onset of seasonal anoestrus in the ewe may be independant of increasing prolactin concentrations and daylength. J. Reprod. Fert. 69, 41-48.

Yuthasastrakosol, P., Palmer, W.M. and Howland, B.E. 1975. Luteinizing hormones, oestrogen and progesterone levels in peripheral serum of anoestrous and cyclic ewes as determined by radioimmunoassay. J. Reprod. Fert. 43, 57-65.

GENETICS AND PHYSIOLOGY OF FOLLICLE RECRUITMENT
AND MATURATION DURING SEASONAL ANOESTRUS

R. Webb, I.K. Gauld

AFRC Animal Breeding Research Organisation,
Dryden Laboratory,
Roslin, Midlothian EH25 9PS U.K.

SUMMARY

Studies using a number of different breeds indicate that mature follicles are present during seasonal anoestrus as well as during the breeding season. Their number is correlated with the ovulation rate of the breed. The mechanisms controlling number of mature follicles are active during seasonal anoestrus as indicated by the response following treatment with oestradiol implants, steroid antisera and unilateral ovariectomy. Details of the control mechanisms have still to be determined, but it is suggested that the control exists at both the hypothalamic/pituitary gland and ovarian levels.

I   INTRODUCTION

Ovulation rate and presumably the patterns of follicular growth vary considerably between breeds of sheep and between strains of sheep within a breed. These genetic differences in ovulation rate and follicular growth provide a powerful tool with which to study the mechanisms controlling ovulation rate. Generally, sheep have clearly defined breeding seasons and periods of seasonal anoestrus. Recently it has been proposed (Webb and Gauld, 1984a) that the primary mechanism controlling the number of mature follicles is functional throughout the various reproductive periods.

The characteristics of mature follicles during the breeding season and during seasonal anoestrus will be outlined. This paper will discuss the development of follicles up to the stage at which they can respond to the stimuli which induces ovulation and corpus luteum formation. The differences in follicular populations and number of mature follicles, in different breeds of sheep, during both the breeding season and seasonal anoestrus will be compared. Finally some of the possible mechanisms involved in the control of follicular growth during seasonal anoestrus will be considered.

II   CHARACTERISTICS OF MATURE GRAAFIAN FOLLICLES

The development and maturation of follicles is correlated with different patterns of steroidogenesis (see review by Webb and Gauld, 1984b). From studies carried out during the breeding season, it would appear that follicles pass through an androgen dominated phase with a limited ability to convert androgens to oestrogens. As follicles grow to the mature Graafian stage, they acquire an ability to convert androgens to oestrogens so that when stimulated by the increased gonadotrophin secretions, oestrogen production increases significantly. This has been demonstrated using a variety of different measurements including follicular fluid concentrations of steroids and oestradiol production of follicles in vitro. As shown in Figure 1 there is a highly significant correlation between in vivo oestradiol production from the ovary and in vitro oestradiol production by follicles removed from the same ovary.

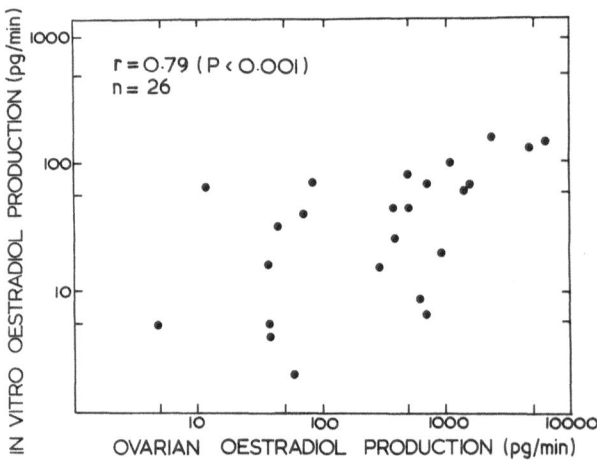

Fig. 1   Correlation between in-vivo ovarian oestradiol production and in-vitro oestradiol production by the three most active follicles removed from the same ovary and incubated individually. n = number of ovaries (Webb and England, unpublished data).

Other important characteristics of these follicles include the presence of significant numbers of LH receptors in the theca and granulosa cells. Furthermore, these follicles are usually found to be among the larger follicles in the ovary and have a full complement of granulosa cells.

From studies carried out in a number of breeds it would appear that healthy follicles possessing these characters can also be found during seasonal anoestrus as well as during the breeding season (see Table 1; McNatty et al., 1984).

TABLE 1   Mean _in vitro_ oestradiol secretion (ng/ml/30 min) by
individual follicles removed from Suffolk ewes at two periods
during the year.

	Mature follicles+	Immature follicles
Breeding season (Follicular phase)	6.7±1.4[a] (n=9)	0.1±0.05 (n=24)
Seasonal anoestrus	1.7±0.5[b] (n=11)	0.1±0.2 (n=34)

+A mature follicle is one producing >500pg/ml/30 min
n = number of follicles; number of mature follicles per ewe
for a=1.8±0.2 and for b=1.8±0.3
(Refs: England and Webb, 1979 and Webb and England, 1982a).

Mature and immature follicles were categorised in two breeds of sheep
(Finnish Landrace and Scottish Blackface), during seasonal anoestrus,
using the presence or absence of an active aromatase enzyme system.   A
follicle having an _in vitro_ oestradiol production of >500pg/ml/hour was
classified as mature.   The mature follicles as well as producing
significant quantities of oestradiol, were larger, had significantly
more granulosa cells and significantly greater numbers of LH receptors
than the immature follicles (Webb, Gauld and Land, 1984).   Interestingly,
no significant differences in testosterone production _in vitro_ were found
between the mature and immature follicles.   This indicates that even
during seasonal anoestrus, as well as during the breeding season, the
developing follicle goes through an androgen dominated phase.

It may be concluded that in a number of breeds of sheep mature
Graafian follicles are present during both the breeding season and
seasonal anoestrus.   If the frequency of LH pulses during seasonal
anoestrus is increased by giving LH or LH-RH, then ovulation and normal
luteal function can be induced (McNatty, Gibb, Dobson and Thurley, 1981;
McNeilly, O'Connell and Baird, 1982; Mc Leod, Haresign and Lamming,
1982a,b).   However, recent evidence has suggested that both progesterone
(McLeod, Haresign and Lamming, 1982b; McLeod and Haresign, 1984) and FSH
(McNeilly, 1984) also play a role in the final maturation of the ovulatory
follicle.

III   THE MATURE GRAAFIAN FOLLICLE POPULATION

Using the characteristics outlined in the previous section the number
of mature follicles in both the luteal and follicular phases of the
oestrous cycle and during seasonal anoestrus have been compared in
different breeds of sheep.   Table 2 illustrates the close correlation
found between mature Graafian follicles and ovulation rate in a number of
breeds of sheep during the follicular phase of the oestrous cycle.

TABLE 2   Ovulation rate and number of mature Graafian follicles
in different breeds of sheep during the follicular phase of the
oestrous cycle.

Breed of sheep	Number of mature follicles/ewe	Ovulation rate/ewe
Suffolk[c]	1.2 ± 0.3^a	1.3 ± 0.1
Finnish Landrace[c]	2.9 ± 0.4^a	2.7 ± 0.2
Booroola Merino (+/+)[d]	1.1 ± 0.1^b	1.0 ± 0.0
Booroola Merino (F/+)[d]	2.1 ± 0.2^b	1.9 ± 0.1
Welsh Mountain[e]	1.3 ± 0.1^b	1.5 ± 0.2

n = 7-35   A follicle was classified as mature (a) with
significant hCG binding to thecal and granulosa cells or (b)
producing >500pg/ml/hr.
(Refs: c. Webb and England, 1982b, d. Webb, Al-Obaidi, Bindon,
Hillard, O'Shea and Piper, unpublished data, e. Webb and Gauld,
unpublished data).

The same relationship also appears to exist during the luteal phase
of the oestrous cycle.   In Suffolk ewes the mean number of follicles
(1.1 ± 0.2), with significant numbers of LH receptors on the theca and
granulosa cells, was similar to the ovulation rate (1.4 ± 0.1) (England
and Webb, 1981).   Furthermore, after treatment of Merino ewes, with hCG,
during the luteal phase the number of induced corpora lutea was the same
as the existing ovulation rate $(1.4\pm0.24$; see review by Scaramuzzi and
Hoskinson, 1984).   During seasonal anoestrus the number of mature
follicles appears to be related to the ovulation rate found in the
breeding season.   This certainly appears to be the case in Scottish
Blackface and Finnish Landrace ewes as demonstrated in Table 3.

These data therefore support the hypothesis that the mechanism
controlling the number of mature follicles is functional during seasonal
anoestrus as well as during the breeding season.   Because of the smaller
variation in hormone concentrations during the period of seasonal
anoestrus compared with the breeding season and especially during the

TABLE 3   Ovulation rate (breeding season) and number of mature
Graafian follicles (seasonal anoestrus) in two breeds of sheep.

	Mature follicles[b] (mean ±S.E.M.)	Ovulation rate (mean ±S.E.M.)
Scottish Blackface	1.2±0.2 (n=18)	1.5±0.1 (n=38)
Finnish Landrace	2.2±0.3 (n=21)	3.1±0.3 (n=15)

n = number of ewes. a) Each value is mean from two experiments,
b) A mature follicle is one producing >500pg/ml/hr.
(Ref: Webb and Gauld, unpublished data).

follicular phase of the oestrous cycle (see review by Baird and McNeilly,
1981; Webb and Gauld, 1984b), the seasonal anoestrous period would appear
to be a more appropriate time in which to investigate the mechanisms
controlling ovulation rate.

IV   CONTROL OF THE NUMBER OF MATURE FOLLICLES

The exact mechanisms controlling the number of mature follicles,
and ultimately ovulation rate, have not yet been elucidated.   Differences
in peripheral gonadotrophin concentrations have been found between breeds
(see review by Bindon and Piper, 1984).   However, the general consensus
is that peripheral hormone concentrations in adult ewes do not fully
explain the differences in ovulation rates found between breeds.   It has
been proposed that there is a variation between breeds in the sensitivity
of the hypothalamus/pituitary gland to the negative feedback effects of
gonadal hormones (see review by Land and Carr, 1979).   This would allow
a difference in ovulation rate without major differences in peripheral
gonadotrophin concentrations.   The following studies were therefore
undertaken to elucidate the possible mechanisms involved in controlling
the number of mature follicles.

To investigate the relationship between ovarian hormones and
ovulation rate different breeds of sheep have been treated with oestradiol
implants.   As shown in Table 4 oestradiol implants cause a marked
reduction in ovulation rate.

Finnish Landrace and Scottish Blackface ewes have been implanted
with oestradiol during seasonal anoestrus.   As in the breeding season

TABLE 4   Ovulation rate in two breeds of sheep following
treatment with an oestradiol implant, during the breeding
season.

Sheep breed	Controls	Treated
Finnish Landrace[a]	3.5±0.4	1.6±0.2
Welsh Mountain[b]	1.5±0.2	0.8±0.1

n = 8-14 ewes. The size of implants used produce peripheral
oestradiol concentrations of approx 5pg/ml (Webb and Baxter,
unpublished data).
(Refs: a) Pathiraja, 1982; b) Webb and Gauld, unpublished data).

the treatment had a significant effect by causing a marked reduction in
the number of mature Graafian follicles (see Table 5).   It is
interesting to note however, that no differences in peripheral FSH
concentrations were found between the control and treated groups in
either breed (Webb, Gauld and Tsonis, unpublished data).

TABLE 5   Effect of oestradiol implants on the number of mature
follicles in two breeds of sheep during seasonal anoestrus.

	Control	Treated
Scottish Blackface (n=11)	0.8±0.2	0.0±0.0
Finnish Landrace (n=12)	2.0±0.4	0.2±0.2

n = number of ewes.   A mature follicle is one secreting >500pg/ml/hr.
Size of implant used have been found to produce peripheral oestradiol
concentrations of approx 2pg/ml (Webb and McBride, unpublished data).
(Ref: Webb, Gauld and Tsonis, unpublished data).

These results cannot differentiate between an effect at the level of the
hypothalamic/pituitary gland or a direct effect of oestradiol at the
ovarian level.

The converse approach is to remove ovarian negative feedback.
Certainly during the breeding season both active and passive immunisation
against steroids leads to a significant increase in ovulation rate and
mean number of lambs born per ewe (see review by Scaramuzzi and Hoskinson,
1984 and Webb, Land, Pathiraja and Morris, 1984).   Seasonally anoestrous

Welsh Mountain ewes have also been passively immunised with a number of steroid antisera. The immunisation treatment did not increase the number of ewes ovulating, but the ovulation rate in the treated ewes that did ovulate, was increased (Land, Fordyce, Gauld, Morris and Webb, 1984). A similar response has been found in Merino ewes actively immunised against androstenedione (Martin, Scaramuzzi and Lindsay, 1981). These treatments therefore separate the mechanisms controlling the number of mature follicles from those which control the final maturation and ovulation. It also suggests that removal of steroid negative feedback leads to an increase in the mature follicle population and then subsequently an increase in ovulation rate. However, as with the oestradiol implant treatment a direct effect at the ovarian level cannot be discounted.

The finely balanced inter-relationship between the ovary, pituitary gland and hypothalamus is further highlighted by the ability of ewes from different breeds to compensate for the removal of an ovary within 3 days (Land, 1973; Findlay and Cumming, 1977). Compensatory follicular hypertrophy can also occur during seasonal anoestrus following unilateral ovariectomy. As shown in Table 6 the number of mature follicles nearly doubled in the remaining ovary compared to the first ovary removed. These results therefore agree with the conclusion of Dufour, Ginther and Casida (1971) using follicular fluid weight to assess follicular activity.

TABLE 6   The number of mature follicles in normal and hypertrophic ovaries of Scottish Blackface sheep during seasonal anoestrus.

	Normal ovary (n=6)	Hypertrophic ovary (n=6)
Follicles producing >500pg/ml/hr oestradiol	7	14
Mean number of mature follicles/ewe ± S.E.M.	1.17±0.40	2.33±0.42

(Ref: Webb, Gauld and Land, 1984).

Findlay and Cumming (1977) found a significant transient increase in FSH following unilateral ovariectomy in the breeding season, but not during seasonal anoestrus. This finding again raises the question of the role of FSH in controlling ovulation rate as follicular hypertrophy does appear to occur during seasonal anoestrus. Furthermore, FSH

concentrations increase following passive immunisation of ewes with
oestrogen antibodies, but not following androgen antiserum, despite both
treatments increasing ovulation rate (Pathiraja et al. 1984). Certainly
infusion of exogenous FSH during the late luteal phase can increase
ovulation rate in Welsh Mountain ewes (Baird, McNeilly, Wallace and Webb,
unpublished observations). Also active immunisation of ewes against a
partially purified preparation of bovine follicular fluid led to an
increased ovulation rate (O'Shea, Cummins, Bindon and Findlay, 1982)
causing an apparent increase in FSH concentrations (Al-Obaidi, Bindon,
Hillard, O'Shea and Piper, personal communication). However, the
partially purified preparation of bovine follicular fluid contains a
large number of proteins so a direct effect at the ovarian level cannot
be discounted. This is supported by recent reports (Cahill, Driancourt
and Findlay, 1984; Cahill, Clarke, Cumming and Findlay, 1984) that ovine
follicular fluid can block a PMSG induced increase in follicular number.

V   CONCLUSIONS

Although oestrous cycles and ovulation do not normally occur during
seasonal anoestrus, follicular development is still taking place.
Follicles having the same characteristics as those found during the
breeding season are present in numbers similar to the ovulation rate.

It is suggested that the mechanisms controlling follicular
development and maturation to the final preovulatory stage are functional
during seasonal anoestrus. These mechanisms appear to be independent
of the final maturation of follicles that occur prior to ovulation,
during the final stages of the follicular phase of the oestrous cycle.

The exact mechanisms controlling the number of follicles which mature
are not clearly defined. Evidence suggests that FSH plays a role, but
the role of LH in these mechanisms is still uncertain (see review by
Bindon and Piper, 1984). Recent evidence suggests that an intra-
ovarian control of follicular growth cannot be discounted. The control
of follicular growth and the number of mature follicles may therefore be
mediated at both the hypothalamus/pituitary gland level and at an ovarian
level. The importance of each site may vary depending on sheep breed
and environmental influences. Finally, the period of seasonal anoestrus
may be the optimum time in which to investigate the control of ovulation
rate and follicular growth.

REFERENCES

Baird, D.T. and McNeilly, A.S. 1981. Gonadotrophic control of follicular development and function during the oestrous cycle of the ewe. J. Reprod. Fert., Suppl. 30, 119-133.

Bindon, B.M. and Piper, L.R. 1984. Endocrine basis of genetic differences in ovine prolificacy. Proc. 10th Int. Cong. Anim. Reprod. and A.I., Univ. Illinois, U.S.A., Vol. IV, Chapt. VI pp. 17-26.

Cahill, L.P., Clarke, I.J., Cummins, J.T. and Findlay, J.K. 1984. Direct inhibition of follicular growth by steroid free follicular fluid. Proc. 16th Ann. Confer. Aust. Soc. Rep. Biol. (Abstr. 21).

Cahill, L.P., Driancourt, M-A, Findlay, J.K. 1984. Inhibition of folliculogenesis by ovine follicular fluid (OFF) in PMSG-treated ewes. 17th Ann. Meeting Soc. Study Reprod., Biol. Reprod. 30, Suppl. 1, 36 (Abstr. 12).

Dufour, J., Ginther, O.J. and Casida, L.E. 1971. Compensatory hypertrophy after unilateral ovariectomy and destruction of follicles in the anoestrous ewe. Proc. Soc. Expt. Biol. Med. 138, 1068-1072.

England, B.G. and Webb, R. 1979. Estradiol production in vitro by Individual follicles from seasonally anoestrous sheep. Proc. 71st Meeting American Soc. Anim. Sci. Abst. No. 380. p. 295.

England, B.G. and Webb, R. 1981. Follicular steroidogenesis and gonadotropin binding to ovine follicles during the estrous cycle. Endocrinology 109, 881-887.

Findlay, J.K. and Cumming, I.A. 1977. The effect of unilateral ovariectomy on plasma gonadotropin levels, estrus and ovulation rate in sheep. Biol. Reprod. 17, 178-183.

Land, R.B. 1973. Ovulation rate in Finn-Dorset sheep following unilateral ovariectomy or chloropromazine treatment at different stages of the oestrous cycle. J. Reprod. Fert., 33, 99-105.

Land, R.B. and Carr, W.R. 1979. Reproduction in domestic animals. In "Genetic Variation in Hormone Systems" (Ed. J.G.M. Shire). (CRC Press, Florida). Vol. 1. Chapt. 5 pp. 89-112.

Land, R.B., Fordyce, M., Gauld, I.K., Morris, B.A. and Webb, R. 1983. Fertility of sheep given antisera to steroids during anoestrus. J. Reprod. Fert., 67, 269-273.

McLeod, B.J. and Haresign, W. 1984. Evidence that progesterone may influence subsequent luteal function in the ewe by modulating pre-ovulatory follicle development. J. Reprod. Fert., 71, 381-386.

McLeod, B.J., Haresign, W. and Lamming, G.E. 1982a. The induction of ovulation and luteal function in seasonally anoestrous ewes treated with small dose multiple injections of GnRH. J. Reprod. Fert., 65, 215-221.

McLeod, B.J., Haresign, W. and Lamming, G.E. 1982b. Response of seasonally anoestrous ewes to small dose multiple injections of GnRH with and without progesterone pre-treatment. J. Reprod. Fert., 65, 223-230.

McNatty, K.P., Gibb, M., Dobson, C. and Thurley, D.C. 1981. Evidence that changes in luteinizing hormone secretion regulate the growth of the preovulatory follicle in the ewe. J. Endocr., 90, 375-398.

McNatty, K.P., Hudson, N.L., Henderson, K.M., Lun, S., Heath, D.A., Gibb, M., Ball, K., McDiarmid, J.M. and Thurley, D.C. 1984. Changes in gonadotrophin secretion and ovarian antral follicular activity in seasonally breeding sheep throughout the year. J. Reprod. Fert., 70, 309-321.

McNeilly, A.S. 1984. Changes in FSH and the pulsatile secretion of LH during the delaying oestrous induced by treatment of ewes with bovine follicular fluid. J. Reprod. Fert., 72, 165-172.

McNeilly, A.S., O'Connell, M. and Baird, D.T. 1982. Induction of ovulation and normal luteal function by pulsed injections of luteinizing hormone in anestrous ewes. Endocrinology, 110, 1292-1299.

Martin, G.B., Scaramuzzi, R.J. and Lindsay, D.R. 1981. Induction of ovulation in seasonally anovular ewes by the introduction of rams: effects of progesterone and active immunization against androstenedione. Aust. J. Biol. Sci., 34, 569-575.

O'Shea, T., Cummins, L.J., Bindon, B.M. and Findlay, J.K. 1982. Increased ovulation rate in ewes vaccinated with an inhibin - enriched fraction from bovine follicular fluid. Proc. Aust. Soc. Reprod. Biol., 14, 85 (Abstr).

Pathiraja, N. 1982. Physiological basis of genetic variation in ovulation rate. Ph.D. Thesis. University of Edinburgh.

Pathiraja, N., Carr, W.R., Fordyce, M., Forster, J., Land, R.B. and Morris, B.A. 1984. Concentration of gonadotrophins in the plasma of sheep given gonadal steroid antisera to raise ovulation rate. J. Reprod. Fert., 72, 93-100.

Scaramuzzi, R.J. and Hoskinson, R.M. 1984. Immunoneutralization of steroid hormones for increasing fecundity. In "Immunological Aspects of Reproduction" (Ed. D.B. Crighton). (Butterworths, London). (in press).

Webb, R. and England, B.G. 1982a. Relationship between LH receptor concentrations in thecal and granulosa cells and in-vivo and in-vitro steroid secretion by ovine fluids during the preovulatory period. J. Reprod. Fert., 66, 169-180.

Webb, R. and England, B.G. 1982b. Identification of the ovulatory follicle in the ewe; associated changes in follicular size, thecal and granulosa cell luteinizing hormone receptors, antral fluid steroids, and circulating hormones during the preovulatory period. Endocrinology, 110, 873-881.

Webb, R. and Gauld, I.K. 1984a. Folliculogenesis in sheep: Control of ovulation rate In "Genetics of Reproduction in the Sheep" (Eds. R.B. Land and D.W. Robinson). (Butterworths, London). (in press).

Webb, R. and Gauld, I.K. 1984b. Final maturation of the preovulatory follicle in the ewe. Extrait du colloque de la Société français pour l'étude de la Fertilité. Periode Péri-ovulatoire. Edition Masson, Paris. pp. 21-31.

Webb, R., Gauld, I.K. and Land, R.B. 1984. Seasonal independence of follicle development in the ewe. Proc. 10th Int. Congr. Anim. Reprod. and A.I. Univ. Illinois, U.S.A. Vol. III, Paper No. 498.

Webb, R., Land, R.B., Pathiraja, N. and Morris, B.A. 1984. Passive immunization against steroid hormones in the female. In " Immunological Aspects of Reproduction" (Ed. D.B. Crighton). (Butterworths, London). Chapt. 26, pp. 475-499.

BREED DIFFERENCES IN THE BREEDING SEASON IN SHEEP

J.F. Quirke*,J.P. Hanrahan**

*Agricultural Institute, Grange, Dunsany, Co. Meath.
**Agricultural Institute, Belclare, Tuam, Co. Galway.

ABSTRACT
    The literature has been surveyed for evidence of genetic
variation in sexual function of rams throughout the year and
in the timing of onset and duration of the breeding season
in female sheep. The results of studies on the breeding
season of several ewe breeds in Ireland are also presented.
During two years first oestrus occurred one month later in
Texel than in Suffolk ewes. First oestrus occurred at the
end of August in Finn-Dorset, mid September in Galway,
Fingalway and Suffolk-crossbreds and during early October in
Scottish Blackface ewes. Oestrous cycles ceased in early
February in Scottish Blackface ewes, in early March in
Galway and Suffolk-crossbreds and during early April in
Fingalways and Finn-Dorsets. The repeatability of date of
onset of first oestrus, date of cessation of oestrous
activity and duration of the anoestrum in consecutive years
was 0.40 ± 0.08, 0.25 ± 0.09 and 0.30 ± 0.09 respectively.

INTRODUCTION

    The sheep is widely known as an animal with marked
seasonality of breeding. It is also well established that
seasonal fluctuation of the photoperiod is the predominant
factor influencing the phasing of reproductive activity of
breeds originating and maintained in the temperate
latitudes(Yeates 1949).The gradual reduction in day length
occuring from mid summer onwards is responsible for the
initiation of oestrous cyclicity while the increase in day
length occuring from mid winter onwards eventually results
in the cessation of oestrous activity. The effects of
altering the photoperiod on reproduction in the sheep and
the endocrine basis of the seasonal anoestrus has been dealt
with elsewhere in these proceedings (Haresign, McLeod and
Webster, 1984; Thimonier and Ortavant, 1984). The purpose
of the present paper is to examine the extent to which
genetic factors contribute to determining the time of onset
and duration of the breeding season in the ewe and on male
reproductive performance.

The degree to which other factors such as age, previous breeding history and the social interactions of the sexes impinge on the manifestation of oestrous and ovulation in the female will also be considered.

REPRODUCTIVE PERFORMANCE OF RAMS

Rams of most breeds are sexually active throughout the year although there is good experimental evidence for seasonal variation in both libido and the quality and quantity of semen produced (Barrell and Lapwood, 1978; Colas, 1979; Islam and Land, 1977; Pepelko and Clegg, 1965; Smyth and Gordon,1967; Thibault et al., 1966). Sexual activity is at a peak in the autumn to coincide with the period of reproductive cyclicity in the ewe and reaches a nadir during the summer months. It has been shown by Schanbacher (1979) that both semen characteristics and reproductive performance of rams during the non-breeding season can be improved by artifically reducing day length. Evidence for increased embryo mortality in ewes inseminated with spermatozoa obtained from rams under an artifically prolonged photoperiod has also been provided by Colas (1983).

The extent to which different ram breeds may vary in semen production, libido and fertility in response to changes in photoperiod has received relatively little attention. A study of rams of the Galway, Suffolk, Cheviot and Dorset Horn breeds in Ireland by Smith & Gordon (1967) revealed little evidence for important differences among the breeds in the seasonal pattern of variation in semen characteristics. Sperm production by a number of French ram breed types during the breeding and non-breeding seasons has been compared using a technique for collecting rete testis fluid (Dacheaux et al., 1981). Production values increased in all breeds during the breeding season and it would appear that Ile de France rams were more susceptible to seasonal effects than many others.

Breed differences in daily sperm production have also been reported by Islam and Land (1977) who observed that Merino rams were less susceptible to changes in the natural photoperiod obtaining in Scotland than those of the Finnish Landrace breed and its crosses with the Dorset and Merino.

Rams of the Finnish Landrace breed are generally considered to be particularly active sexually and evidence for this has been provided in a number of studies. In the comparison of Finnish Landrace and Suffolk rams conducted by Shanbacher and Lunstra (1976), in the United States, seasonal variation in sexual activity was apparent in both breeds but the Finnish Landrace rams were more active than the Suffolks during all months of the year. A similar result was obtained when Finnish Landrace rams were compared with Scottish Blackface (Land, 1970) and with Tasmanian Merino rams (Islam and Land, 1977; Land and Sales, 1977) in Britain.

THE MALE EFFECT

The introduction of rams during the transitional phase between the anoestrous and oestrous seasons following a preconditioning period of isolation from males can stimulate ovulation in many ewes within two to three days (Knight, Peterson and Payne, 1978; Oldham, Martin and Knight, 1979). Oestrus is not usually associated with the ram-induced ovulation but the corpus luteum formed can provide the progesterone priming necessary for oestrus to occur at the subsequent ovulation. Premature regression of the ram-induced corpora lutea followed by a second ovulation after 4-8 days may occur in approximately 50% of ewes (Knight,Tervit and Fairclough, 1981; Oldham and Martin,1979). This accounts for the two distinct peaks of oestrous activity following the initial contact with rams, the first at 18-19 days and the second approximately 6 days later (Fairnie, 1976; Lyle and Hunter, 1967; Schinickel, 1954). The ram stimulated ovulation results from an apparently normal pre-ovulatory LH surge (Martin et al,1980; Oldham, Martin and Knight,1979) in response to the presence of pheromones in the body secretions of the rams (Knight and Lynch,1980 a,b).

Evidence for genetic variation in the effectivness of rams to provoke an ovulatory response in the ewe and advance the timing of first oestrus has been provided by some studies in New Zealand. Romney ewes joined with Dorset rams exhibited first oestrus significantly earlier than those exposed to Merino rams or to rams of their own breed (Meyer,1979; Tervit, Havik and Smith, 1977). The reasons for the superiority of the Dorset rams however are not known. Testosterone evidently has some role in the mediation of the ram effect since it has been shown that androgen treated wethers and ewes are as effective as vasectomised rams in inducing ovulation and cyclicity in anoestrous ewes (Croker, Butler, Johns and McColm, 1982). The substantial differences between Dorset and Romney rams in plasma testosterone levels, however, do not reflect the effectiveness of the breeds in inducing the ram effect in Romney ewes (Tervit and Peterson,1978).

The proportion of ewes which ovulate in response to introduction of rams varies widely in the literature but there is a paucity of comparative data for ewe breeds. It would appear that breeds such as the Merino and Prealpes du Sud are capable of responding at almost any time of the year provided they are in anoestrus whereas those with a deeper anoestrum such as the Romney, appear to be susceptible to the ram effect only during a limited period immediately prior to the start of the spontaneous breeding season (Edgar and Bilkey, 1963; Oldham and Cognie,1980).Meyer (1979) failed to detect differences between the Romney and a range of crosses with the Merino and various imported breeds in New Zealand.

ONSET AND DURATION OF THE BREEDING SEASON IN EWE LAMBS

Age has an important bearing on the time of onset and duration of the breeding season in sheep. Thus ewe lambs commence cycling later and have a shorter breeding season than yearling and older ewes (Hafez,1952).

The time of onset of first oestrus in lambs is subject to a wide range of environmental influences, particularly nutrition and photoperiod, some of which may interact with genetic effects (Dyrmundsson,1983; Land,1978).Ideally, therefore, breed types should be compared under local conditions of husbandry and management. Dyrmundsson (1973) has reviewed and tabulated the published values for pubertal traits for a large number of ewe lamb breeds. Much of the information on genetic variation in the onset of puberty in lambs accumulated since then has been concerned with the effects of crossing the sexually precocious Finnish Landrace on domestic breeds in many counries (Cedillo, Hohenboken and Drummond, 1977; Dickerson and Laster, 1975; Land, Russel and Donald, 1974; Quirke, 1978; 1979; Quirke and Gosling, 1979). These studies show that the Finnish Landrace breed transmits earliness of sexual maturity to female progeny in a wide range of breed crosses. Land (1978) has demonstrated considerable heterosis for this trait and has also shown parallel effects on the sexual maturity of male lambs.

Although there is an abundance of data on the timimg of commencement of cyclicity in numerous ewe lamb breeds information on the duration of the breeding season in very young ewes is much more limited. Hafez (1952) compared the length of the season for Suffolk, Romney and Scottish Blackface lambs in Britain and obtained estimates of 70,38 and 34 days respectively. A similar range in duration of the season was observed in a comparison of the Hampshire, Columbia and Rambouillet breeds in the United States (Foote, Sefidbahkt and Madsen, 1970). In addition to their propensity to display oestrus earlier than many other breed types Finnish Landrace ewe lambs can, evidently, also continue to experience estrous cycles for an extended period during their first breeding season. This is shown clearly in the results of a trial conducted in our Institute some years ago (Quirke,1978; Table 1).

34

Table 1   Mean dates of onset and cessation of oestrous activity and number of oestrous periods during the breeding season in Finnish Landrace and Galway ewe lambs. (Adapted from Quirke, 1978).

	Galway	Finnish Landrace
Number of ewe lambs	157	131
Number attained puberty (%)	71(45)	113(86)
Mean date of the first observed oestrus	December 5	November 30
Mean date of the last observed oestrus	December 22	January 18
Number of oestrous periods	1.7	2.9

Although the Finnish Landrace ewe lambs in this Irish study did not attain puberty any earlier than those of the Galway breed they nevertheless continued to cycle until the second half of January whereas the Galway lambs ceased cyclicity in late December. A similar result was obtained by Land et al.(1973) who observed that ewe lambs of the prolific Romanov breed had a significantly extended breeding season compared with those of the non-prolific Solognote breed although both breeds attained puberty at a similar age and time of year.

ONSET AND DURATION OF THE BREEDING SEASON IN MATURE EWES

The timing of onset and duration of the breeding season has been established for many breeds and crosses in particular environments during the past 40 years. Since the present discussion is concerned with genetic variation in these traits consideration will be given here only to reports dealing with the reactions of different breeds and crosses to the same environmental conditions.

In Britian the Dorset Horn breed commences cycling during late July and is evidently the earliest breed to do so in that country (Hafez, 1952). The mean dates of onset of first oestrus in Suffolk,Romney and Border Leicester ewes in the same study occured in early October - some two to three weeks ahead of the Scottish Blackface and Welsh Mountain breeds which, unlike the other breeds are normally maintained in the less favourable hill regions. Oestrous activity ceased during the month of February in the Border Leicester and Hill breeds but continued into early March for the Dorset Horn and other lowland breeds. The ability of ewes of the Dorset Horn breed to commence cycling earlier than a range of other domestic breeds in New Zealand has been demonstrated recently (Kelly,Allison and Schakell, 1976). Earliness of initation of cyclicity however is not necessarily associated with a particularly late cessation of oestrous activity since in both the British and New Zealand studies the Dorset Horn ewes ceased cycling earlier than many of the other breeds studied (Hafez, 1952; Kelly,Allison and Schakell,1976). In Ireland Suffolk ewes have been shown to commence oestrous activity one month earlier than contemporary Texel ewes (Hanrahan, 1982; Table 2).

Table 2   Median date of first oestrus in Suffolk and
                 Texel ewes. (Adapted from Hanrahan, 1982).

Year of observation	Ewe Breed	Number of Ewes	Median date of first oestrus
1979	Suffolk	52	August 29
	Texel	40	September 25
1980	Suffolk	44	August 25
	Texel	27	September 26

Differences among breeds in the time of onset and duration of the breeding season have also been reported from the United States and Canada (Dufour,1974; Lax et al., 1979) and it would appear that the breeds studied have a well defined period of sexual inactivity.

A comparative, and previously unpublished, study of the breeding season in a number of ewe breeds in Ireland has recently been completed. The experiment commenced in November,1979 when vasectomized rams were joined with mature Galway, Suffolk crossbred, Fingalway and Finn-Dorset ewes which were maintained as a single flock outdoor until November,1981. Beginning in September, 1980 a group of adult Scottish Blackface ewes was also included in the study. The ewes were inspected daily for evidence of oestrus throughout the entire two year period of the experiment.

The ovaries of all ewes were inspected for the presence of luteal structures during the first week of July,1981 using an endoscope in order to determine if any ewes were experiencing silent ovulation during the anoestrum.

Table 3: Mean dates of onset and cessation of the breeding season and duration of the anoestrum in five ewe breed types in Ireland.

	Breed type				
	Galway	Suffolk Cross	Fin- Galway	Finn- Dorset	Scottish Blackface
No. of ewes	27	33	25	28	22
Date of first oestrus	13 Sept.	13 Sept.	11 Sept.	30 Aug.	7 Oct.
Date of final oestrus	6 Mar.	5 Mar.	4 Apr.	5 Apr.	2 Feb.
Length of the Anoestrum (days)	191	190	160	147	247

The main results of the experiment are summarised in table 3 and the values given for the various parameters represent the mean of the observations taken over two seasons for all breeds with the exception of the Scottish Blackface which was present for only one season and will be considered separately. The mean date of onset of the breeding season was similar for all breeds with the exception of the Finn-Dorset which commenced cycling approximately two weeks earlier than the other breeds. The time of onset of the season was similar in both years for all breeds and the repeatability of this parameter, estimated on a pooled within breed basis, was 0.40_0.08. This value is similar to the estimates reported by Hanrahan (1982) and to the repeatability estimate of date of first lambing reported in a Canadian study by Fahmy (1982). All breeds ceased cycling 3 weeks earlier on average in 1982 than during the previous year. The repeatability of the date of the last observed oestrus and the duration of the anoestrum was 0.25+0.09 and 0.30+0.09 respectively. These results suggest the existence of within breed genetic variation in components of the breeding season. This interpretation is consistent with the evidence presented by Land (1982) and Purser (1971) for considerable genetic variation within breeds in the date of onset of the breeding season.

A particularly striking feature of the results was the extent to which the ewes with Finnish Landrace ancestry continued to experience oestrus into the month of April whereas cycles had ceased a month earlier in the Galway and Suffolk cross ewes. The extension of the breeding season in this direction in the Finn-cross ewes in the present work is consistent with the observation by Wheeler and Land (1977) that purebred Finnish Landrace ewes in Scotland continue experiencing oestrous cycles until early in the month of May and with the results of the study of Finnish Landrace and Galway ewe lambs referred to earlier (Table 1).

On the occassions when the Scottish Blackface ewes were present in the flock they commenced cycling later and ceased cycling earlier than any of the other breeds, a result in agreement with the reports for this breed cited earlier.

It was not pōssible to judge the extent to which ewes may have experienced ovarian cycles without expression of behavioural oestrous during the anoestrous season. It is certain however that none of the breeds continued to have silent ovulations throughout the entire anoestrum as luteal structures were not detected in any sheep at the ovarian inspection conducted during July 1981.

Ovulation unaccompanied by oestrus is a widespread phenomenon in sheep. Studies in this area have been facilitated by the development of the technique of endoscopy which permits the ovaries of individual ewes to be viewed repeatedly without detriment to their reproductive function. Thimonier and Mauleon (1969) were the first to employ the technique for this purpose in a comparative study of the Prealpes du Sud and Ile de France breeds and showed that although all individuals had a period of anoestrus a proportion ovulated throughout the year. The technique has been used in several comparative studies since then to examine the relationships between the incidence of oestrus and ovulation throughout the year (Bindon and Piper, 1976; Kelly, Allison and Shackell, 1976; Lahlou-Kassi, 1980; Land et al., 1973; Wheeler and Land,1977). The results of these studies provide evidence for considerable variation among breeds in the duration and level of ovarian activity during the year and in the degree to which ovulation can occur unassociated with the symptoms of behavioural oestrus.

RESUMPTION OF OESTRUS POSTPARTUM

Many factors influence the interval to the resumption of cyclicity postpartum and the literature on this subject has been reviewed by Hunter (1968) and Van Niekerk (1976). Of these the most important are the duration of lactation and season of lambing. Lactation delays the resumption of oestrous activity (Gould and Whiteman, 1973) and ewes which lamb shortly before or during the anoestrous period have a prolonged interval to first oestrous (Dufour, 1975). Breed differences in the timing of the onset of seasonal anoestrus can confound the evaluation of breed effects on the interval to first postpartum oestrus. Quirke et al. (1983) failed to detect significant differences among fall-lambing Dorset, Rambouillet and Finnish Landrace ewes in the postpartum interval to first oestrus whereas Land (1971) observed significant differences for this trait among ewes of the Finnish Landrace, Dorset and Finn-Dorset breeds. In the latter study, however, parturition occurred closer to the end of the breeding season for the Dorset than for the Finnish Landrace breed.

REFERENCES

Barrell, G.K. and Lapwood, K.R. Seasonality of semen production and plasma luteinizing hormone, testosterone and prolactin levels in Romney, Merino and Polled Dorset rams. Anim. Reprod. Sci., 1, 213-228.

Bindon, B.M. and Piper L.R. 1976. Assessment of new and traditional techniques of selection for reproduction rate. In Proc. Int. Sheep Breeding Congr. Edited by G.J. Tomes, D.E. Robertson & R.J. Lightfoot. pp 357-371. Western Australian Institute Technology Press.

Cedillo, R.M., Hohenboken, W. and Drummond, J. 1977. Genetic and environmental effects on age at first oestrus and on wool and lamb production of crossbred ewe lambs. J. Anim. Sci., 44, 948-957.

Colas, G. 1979. Fertility in the ewe after artifical insemination with fresh and frozen semen at the induced oestrus and influence of the photoperiod on the semen quality of the ram. Livest. Prod. Sci., 6, 153-166.

Colas, G. 1983. Factors effecting the quality of ram semen. In: Sheep Breeding. Edited by W. Haresign. Chapter 23, pp 453-466. Butterworths, London.

Croker, K.P., Butler, L.G., Johns, M.A. and McColm, S.C.1982. Induction of ovulation and cyclic activity in anoestrous ewes with testosterone treated wethers and ewes. Theriogenology, 17, 349-354.

Dacheaux, J.L., Pisselet, C., Blanc, M.R.,
        Hochereau-de-Reviers, M.T. and Courot, M. 1981.
        Seasonal variations in rete testis fluid secretion
        and sperm production in different breeds of ram. J.
        Reprod. Fert., 61, 363-371.
Dickerson, G.E. and Laster, D.B. 1975. Breed, heterosis and
        environmental influences on growth and puberty in
        ewe lambs. J. Anim. Sci., 41, 1-9.
Dufour, J.J. 1974. The duration of the breeding season of
        four breeds of sheep. Can. J. Anim. Sci., 54,
        389-392.
Dufour, J.J. 1975. Effects of seasons on postpartum
        characteristics of sheep being selected for
        year-round breeding and on puberty of their female
        progeny. Can. J. Anim. Sci., 55, 487-492.
Dyrmundsson, O.R. 1973. Puberty and early reproductive
        performance in sheep. 1. Ewe lambs. Anim. Breed.
        Abstr., 41, 273-289.
Dyrmundsson, O.R. 1983. The influence of environmental
        factors on the attainment of puberty in ewe lambs.
        In: Sheep Breeding. Edited by W. Haresign. Chapter
        20, pp 393-408.
Edgar, D.G. & Bilkey, D.A. 1963. The influence of rams on
        the onset of the breeding season. Proc. N.Z. Soc.
        Anim. Prod. 23, 79-87.
Fahmy M.H. 1982. Genetic parameters of date of lambing in
        DLS sheep. In: Proc. Wld. Congr. Sheep & Beef
        Cattle Breeding. Vol. 1. pp 401-404. Edited by R.A.
        Barton & W.C. Smith. Dunmore Press Ltd., New
        Zealand.
Fairnie, I.J. 1976. Organisation of artificial breeding
        programs of Sheep in Western Australia. In: Proc.
        Int. Sheep Breeding Congr. Editors G.T. Tomes, D.E.
        Robertson & R.J. Lightfood. Western Australian
        Institute Technology Press, pp 500-508.
Foote, W.C., Sefidbakht, N. and Madsen, M.A. 1970. Puberal
        estrus and ovulation and subsequent estrous cycle
        patterns in the ewe. J. Anim. Sci., 30, 86-90.
Gould, M.B. and Whiteman, J.V. 1973. Postpartum
        reproductive performance of early weaned spring
        lambing ewes. J. Anim. Sci., 36, 1041-1043.
Hafez, E.S.E. 1952. Studies on the breeding season and
        reproduction of the ewe. J. Agric. Sci. Camb., 42,
        189-265.
Haresign, W., McLeod, B.J. and Webster, G.M. 1984.
        Endocrine basis of seasonal anoestrus in Sheep.
        In: Endocrine causes of seasonal and lactational
        anoestrus in farm animals. EC-Seminar, Mariensee,
        October 2-3, 1984.
Hanrahan, J.P. 1982. Repeatability of date of first
        oestrus. Animal Production Research Report, An
        Foras Taluntais, Dublin. Page 92.
Hunter, G.L. 1968. Increasing the frequency of pregnancy in
        sheep. I. Some factors affecting re-breeding during

the postpartum period. Anim. Breed. Abstr., 36,
      347-378.
Islam, A.B.M.M. and Land, R.B. 1977. Seasonal variation in
      testis diameter and sperm output of rams of
      different prolificacy. Anim. Prod., 25, 311-317.
Kelly, R.W., Allison, A.J.A. and Shackell, G.H. 1976.
      Seasonal variation in oestrous and ovarian activity
      of five breeds of ewes in Otago. N.Z. Jl. Exp.
      Agric., 4, 209-214.
Knight, T.W. and Lynch, P.R. 1980a . The pheromone from
      rams that stimulates ovulation in the ewe. Proc.
      Aust. Soc. Anim. Prod., 13, 74-76.
Knight, T.W. and Lynch, P.R. 1980 b. Source of ram
      pheromones that stimulate ovulation in the ewe.
      Anim. Reprod. Sci., 3, 133-136.
Knight, T.W., Peterson, A.J. and Payne, E. 1978. The
      ovarian and hormonal response of the ewe to
      stimulation by the ram early in the breeding
      season. Theriogenology, 10, 343-353.
Knight, T.W., Tervit, H.R. and Fairclough, R.J. 1981.
      Corpus luterm function in ewes stimulated by rams.
      Theriogenology, 15, 183-190.
Lahlou-Kassi, A. 1980. Seasonal variation in oestrus and
      ovarian activity of two Moroccan breeds: D'Man and
      Timhadite. Proc. 9th Int. Cong. Anim. Reprod. &
      A.I., Madrid, Vol IV, pp 186 -189.
Land,R.B. 1970. The mating behaviour and semen
      characteristics of Finnish Landrace and Scottish
      Blackface rams. Anim. Prod., 12, 551-560.
Land, R.B. 1971. The incidence of oestrus during lactation
      in Finnish Landrace, Dorset Horn and Finn-Dorset
      sheep. Anim. Prod., 24, 345-352.
Land, R.B. 1978. Reproduction in young sheep: some genetic
      and environmental sources of variation. J. Reprod.
      Fert., 52, 427-436.
Land, R.B. 1982. Indications of reproductive potential.
      In: Proc. Wld. Congr. Sheep & Beef Cattle Breeding.
      Vol. 1. pp 365-373. Edited by R.A. Barton & W.C.
      Smith. Dunmore Press Ltd., New Zealand.
Land, R.B. and Sales, D.I. 1977. Mating behaviour and
      testis growth of Finnish Landrace, Tasmanian Merino
      and crossbred rams. Anim. Prod., 24, 83-90.
Land, R.B., Pelletier, J., Thimonier, J. and Mauleon, P.
      1973. A quantitative study of genetic differences
      in the incidence of oestrus, ovulation and plasma
      luteinizing hormone concentration in the sheep. J.
      Endocr., 58, 305-317.
Land, R.B., Russel, W.S. and Donald, H.P. 1974. The litter
      size and fertility of Finnish Landrace and

Tasmanian Merino Sheep and their reciprocal crosses. Anim. Prod. 18, 265-271.

Lax, J., French, L.R., Chapman, A.B., Pope, A.L. and Casida, L.E. 1979. Length of breeding season for eight breed groups of sheep in Wisconsin. J. Anim. Sci., 49, 939-942.

Lyle, A.D. and Hunter, G.L. 1967. Teasing groups of ewes at staggered intervals as a means of levelling the ram mating load and flock lambing rate. S. Afr. J. Agric. Sci., 10, 597-508.

Martin, G.B., Cognie, Y., Gayerie, F., Oldham, C., Poindron, P., Scaramuzzi, R.J. & Thiery, J.C. 1980. The hormonal responses to teasing. Proc. Aust. Soc. Anim. Prod., 13, 77-79.

Meyer, H.H. 1979. Ewe and teaser breed effects on reproductive behaviour and performance. Proc. N.Z. Soc. Anim. Prod., 39, 68-76.

Oldham, C.M. and Martin, G.S. 1979. Stimulation of seasonally anovular Merino ewes by rams. 11. Premature regression of ram-induced corpora lutea. Anim. Reprod. Sci., 1, 291-295.

Oldham, C. Martin, G.S. & Knight, T.W. 1979. Stimulation of seasonally anovular Merino ewes by Rams. 1. Time from introduction of rams to the pre-ovulatory LH surge and ovulation. Anim. Reprod. Sci., 1, 283-290.

Oldham, C. and Cognie, Y. 1980. Do ewes continue to cycle after teasing ?. Proc. Aust. Soc. Anim. Prod., 13, 82-86.

Pepelko, W.E. and Clegg, M.T. 1965. Influence of season of the year upon patterns of sexual behaviour in male sheep. J. Anim. Sci., 24, 633-637.

Purser, A.F. 1972. Variation in date of first oestrus among Welsh Mountain ewes. Proc. British Soc. Anim. Prod., 1 (New Series), p 133.

Quirke, J.F., 1978. Onset of puberty and oestrous activity in Galway, Finnish Landrace and Finn-cross ewe lambs. Ir. J. Agric. Res., 17, 15-23.

Quirke, J.F. 1979. Effect of bodyweight on the attainment of puberty in Galway and Fingalway female lambs. Anim. Prod. 28, 297-307.

Quirke, J.F. & Gosling, J.P. 1979. Prepubertal plasma luteinizing hormone and progesterone concentrations during the oestrous cycle and early pregnancy in Galway and Fingalway female lambs. Anim. Prod., 28, 1-12.

Quirke, J.F., Stabenfeldt, G.H. and Bradford, G.E. 1983. Resumption of ovarian function in autumn lambing Dorset, Rambouillet and Finnish Landrace ewes. Theriogenology, 19, 243-248.

Schanbacher, B.D. 1979. Increased lamb production with rams

exposed to short day lengths during the nonbreeding season. J. Anim. Sci., 49, 927-932.

Schanbacher and Lunstra, D.D. 1976. Seasonal changes in sexual activity and serum levels of LH and testosterone in Finnish Landrace and Suffolk rams. J. Anim. Sci., 43, 644-650.

Schinckel, P.G. 1954. The effect of the ram on the incidence and occurrence of oestrus in ewes. Aust. Vet. J., 30, 189-195.

Smyth, P. and Gordon, I. 1967. Seasonal and breed variations in the semen characteristics of rams in Ireland. Irish Vet. J., 21, 222-233.

Tervit, H.R., Havik, P.G. & Smith, J.F. 1977. Effect of breed of ram on the onset of the breeding season in Romney ewes. Proc. N.Z. Soc. Anim. Prod. 37, 143-148.

Tervit, H.R. & Peterson, A.J. 1978. Testosterone levels in Dorset and Romney rams and the effectiveness of these breeds in stimulating early onset of oestrus in Romney ewes. Theriogenology, 9, 271-277.

Thimonier, J. and Mauleon, P. 1969. (Seasonal variations in oestrous behaviour and ovarian and pituitary activities in the ewe.) Annls. Biol. Anim. Biochim. Biophys., 9, 233-250.

Thimonier, J. and Ortavant, R. 1984. Control of reproduction in the ewe by light. In: Endocrine causes of seasonal and lactational anestrus in farm animals. EC-Seminar, Mariensee, October 2-3, 1984.

Thibault, C., Courot, M., Martinet, L., Mauleon, P., Du Mesnil Du Buisson, F., Ortavant, R., Pelletier, J., Signoret, J.P. 1966. Regulation of breeding season and estrous cycles by light and external stimuli in some mammals. J. Anim. Sci., 25, Suppl., 119-139.

Van Niekerk, C.H. 1976. Limitations to female reproductive efficiency. In: Proc. Int. Sheep Breeding Congress. Edited by G.J. Tomes, D.E. Robertson & R.J. Lightfoot. pp 299-309. Western Australian Institute of Technology.

Wheeler, A.G. and Land, R.B. 1977. Seasonal variation in oestrus and ovarian activity of Finnish Landrace, Tasmanian Merino and Scottish Blackface ewes. Anim. Prod., 24, 363-376.

Yeates, N.T.M. 1949. The breeding season of the sheep with particular reference to its modification by means of artifical light. J. Agric. Sci. Camb., 39, 1-42.

44

LIGHT CONTROL OF REPRODUCTION IN THE EWE

J. Thimonier, R. Ortavant

Institut National de la Recherche Agronomique
Station de Physiologie de la Reproduction
Nouzilly, 37380 Monnaie, France

ABSTRACT

In mid-latitudes, ewes have a clear seasonal pattern of sexual activity which generally begins after the summer solstice and ends after the winter solstice. Photoperiod is a strong entraining agent of reproductive activity, as demonstrated either by the reversal of annual photoperiodic variations, by rhythms reproducing normal photoperiodic variations within a period of less than 1 year, or by the alternation of long and short days every 3 or 4 months. Each of these treatments modifies the period of sexual activity. It is thus possible to control the periods of reproduction in ewes by putting the animals in light-proof buildings.

Recent studies using split photoperiodic treatments have demonstrated the feasibility of controlling periods of ovarian activity in ewes. Priming cyclic animals whith a 1-hour light pulse, given during the dark period 16 to 17 hours after dawn, and changing the time of that light pulse to between 8 and 12 hours after dawn induces ovarian activity within 2 months after the change.

This makes it possible to control reproduction in ewes by giving only extra light and without having to use light-proof buildings.

INTRODUCTION

In spite of centuries of domestication, ewes of all breeds in mid-latitudes have a well-marked seasonal pattern in cyclic oestrous activity. The sexual season generally begins after the summer solstice (end of summer-beginning of autumn) and, if the ewes remain non-pregnant, the sexual season ends after the winter solstice (end of winter-beginning of spring). There are large variations between breeds as to the onset, end and duration of the sexual season (Thimonier and Gauthier, 1984; Fig. 1) and between the ewes of a given breed (Thimonier and Mauléon, 1969).

Ovarian activity follows a pattern similar to oestrous activity. However, the first and/or second period of ovulation of the season is silent and so is the last ovulatory one. In some breeds silent ovulations occurring during mid-anoestrus are not related to either the end or the onset of the sexual season (Thimonier and Mauléon, 1969; Land et al., 1973). However, deep anoestrus is a general occurrence in all breeds studied until now, at least in France.

Thus, even in sheep which have been domesticated for a long time, factors are involved in the stimulation and/or inhibition of ewe sexual activity. Seasonality is not limited to the females. Libido fluctuates widely in rams, being lowest in spring and highest by the end of summer and autumn (Rouger, 1974). Daily sperm production/testis is minimal from February to May and maximal from August to September (Dacheux et al., 1981) and varies from 1 to nearly 5×10^9 spermatozoa/testis/day in the Ile-de-France breed.

In mid-latitudes, seasonal variations in temperature and daylength are the two main factors having either a direct effect on the animals or an indirect one through forage availability and quality.

High temperatures have been demonstrated to delay the onset of the sexual season (Dutt, 1960) and cool temperatures to advance it (Godley et al., 1966) but temperature does not appear to be the main factor involved in the control of seasonality in ewes. Under an equatorial light regime, temperature variations, when reversed, do not modify the normal rhythm of ewe reproduction (Wodzicka-Tomaszewska et al., 1967).

Photoperiod appears to be the main cue in the control of seasonality in ewes.

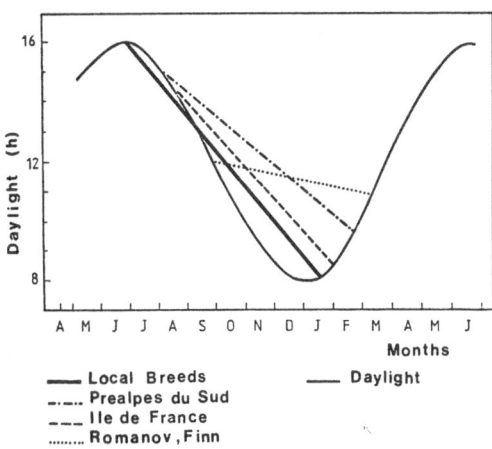

Fig. 1   The sexual season of some breeds used in France in relation to the curve of daylight hours (from Thimonier and Gauthier, 1984).

PHOTOPERIOD AND REPRODUCTION IN EWES

Since Bissonnette (1941) first experimentally demonstrated the role
of light in the control of cyclic oestrous activity in goats, three main
types of experiments have been carried out:
- reversal of the annual photoperiodic cycle;
- rhythms reproducing the normal photoperiodic variation within a
  period of less than one year (6 to 8 months);
- alternation of long and short days every 3 or 4 months.

When English crossbred ewes were submitted to a reversed annual cycle
from October onwards, they had a new sexual season in May, while the
controls maintained under natural photoperiod had a normal sexual season
in August-September (Yeates, 1949). An almost complete reversal of the
sexual season under reversed photoperiod was further confirmed in
Southdown ewes by Thwaites (1965).

Limousine and Préalpes du Sud ewes under a semestrial rhythm had two
periods of sexual activity, starting just before the increase of daylength
and ending when daylight began to decrease (Mauléon and Rougeot, 1962;
Fig. 2). The Corsican moufflon (Ovis ammon musimon) when submitted to the
same photoperiodic treatment reproduces twice a year, mating occurring in
the longest photoperiod (Rougeot, 1969).

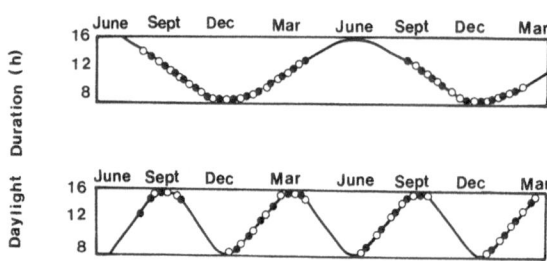

Fig. 2   Appearance of oestrus for individual ewes (,) under annual
(upper part) or semestrial (lower part) rhythms of daylight duration.
Upper part: Under artificial annual variations of daylight oestrous
       activity is confined to one period per year.
Lower part: Under semestrial rhythms with the same amplitude of daylight
       variation, ewes exhibit two periods of sexual activity each year (From
       Mauléon and Rougeot, 1962).

Finally, by alternating long (16 h) and short (8 h) daylength every 3 or 4 months, Legan and Karsch (1980) demonstrated that periods of anoestrus and sexual activity alternate every 3 or 4 months. In Suffolk ewes, sexual activity appears 50 days after the shift to short days and stops with about the same lag after the shift to long days. In an experiment in progress, similar results have been obtained in Ile de France ewes (Thimonier, work in progress).

The role of photoperiod in the control of reproductive activity has been futher demonstrated by blinding ewes (Legan and Karsch, 1983). These ewes lose the ability to respond to light treatment, thus showing the importance of retinal photoreceptors. However, the sexual season of these animals maintains a circannual rhythmicity (Clegg et al., 1965; Legan and Karsch, 1983), indicating the existence in ewes of an internal endogenous circannual rhythm or of other cues in the absence of the main one which is photoperiod.

Endocrinological studies have clearly demonstrated the role of pituitary hormones (FSH and LH) in the control of reproduction. In ewes (Legan and Karsch, 1983) as well as in rams (Pelletier, 1971), the steroid feedback on the hypothalamo-pituitary axis is modified by light. The pineal is involved in this photoperiodic modification of the steroid feedback. Very recently the circadian rhythm of the secretion of melatonin, the pineal hormone, under artificial alternating short and long days has been shown to determine the reproductive response to photoperiod (Karsch et al., 1984). This may offer new possibilities for the control of ewe reproductive activity in the near future.

PRATICAL IMPLICATIONS

Photoperiodic control of reproduction in ewes is feasible and has been carried out successfully.

In all the treatments, daylength was first increased naturally or artificially (14 to 20 h of light per day) and then progressively or abruptly decreased. Sexual activity was induced during decreasing daylength or short days.

Depending on the experiment, the period of the photoperiodic treatment varied from 6 to 8 months (Evans and Robinson, 1980; Hackett, 1982; Heaney et al., 1980; Robinson et al., 1975; Vesely, 1975).

Using such photoperiodic treatments in conjunction with progestagen-PMSG treatments, Evans and Robinson (1980) and Robinson et al. (1975) studying prolific crossbred ewes obtained more than 3 lambs per ewe per year.

However, these photoperiodic treatments involve light-proof buildings. If such investment is feasible for controlling the sexual activity and sperm production of high-genetic value rams used on large numbers of ewes by artificial insemination, farmers in mid-latitudes cannot afford to invest a lot of money in light-proof buildings for this purpose in ewes. It is therefore necessary to find a cheaper solution.

SKELETON LIGHT TREATMENT AND CONTROL OF REPRODUCTION

The existence of a photosensitive phase for LH, testosterone, FSH, prolactin secretions and testis growth has been demonstrated by Ortavant (1977) and confirmed by several experiments (Pelletier et al., 1981; Ravault et al., 1981; Terqui et al., 1984).

To our knowledge, no study until our work has been conducted on ewes using so-called scotophase scan experiments.

In the following studies, the effect of light treatments was estimated according to ovarian activity (or inactivity): plasma progesterone levels were measured in blood samples taken twice weekly to avoid having to detect oestrus with rams and thus induce ovarian activity (ram effect).

A. Ewes preconditionned in long photoperiods were subjected for 15 months to normal variations of daylength (47° N latitude) or to 8 h of illumination in two fractions of 7 and 1-hour duration each, respectively. The beginning of the 7-hour fraction was considered to be subjective dawn and the 1-hour light fraction was given at various times after dawn (8th, 11th, 14th, 17th or 20th hour) according to the experimental group (Ortavant, 1977; Fig. 3).

The results were the following (Table 1) (Thimonier et al., 1978):

- regardless of the position of the 1-hour light fraction, treatment did not affect the onset of the first period of ovarian activity;

- the length of the first period of ovarian activity was significantly shorter in light-treated ewes than in those submitted to normal variations in daylength: the first period of ovarian activity ended earlier than in the controls (mid-November vs beginning of February),

- the second period of ovarian activity also began earlier (P < 0.05) in all light-treated ewes. However, this second period of ovarian activity was characterized by abnormal duration and irregular seccession of cycles.

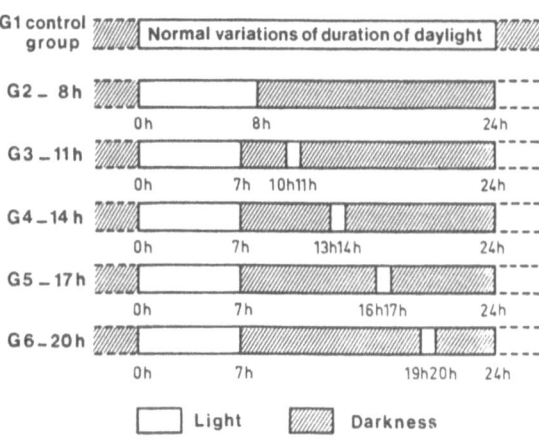

Fig. 3   Experimental procedure to demonstrate a photosensitive phase.

As a result, deep anoestrus (ovarian inactivity) was also significantly shorter for 4 out of 5 light-treated groups.

TABLE 1   Ovarian activity in light-treated ewes.

Light treatment	Onset of 1st period (date)	Duration (days)	End of 1st period (date)	Duration of ovarian inactivity (days)	Onset of 2nd period (date)
G1	5-8 ± 6.7	182 ± 11.0	3-2 ± 5.3	207 ± 5.9	28-8 ± 5.0
G2 8 hrs	4-8 ± 2.6	99 ± 5.7	11-11 ± 6.5	206 ± 7.1	4-6 ± 10.1
G3 10-11	2-8 ± 3.1	93 ± 5.6	2-11 ± 4.6	161 ± 6.9	12-4 ± 7.7*
G4 13-14	30-7 ± 3.6	118 ± 10.6	24-11 ± 10.7	168 ± 12.6	12-5 ± 8.7
G5 16-17	12-8 ± 7.8	107 ± 15.4	29-11 ± 9.7	158 ± 12.9	6-5 ± 12.2
G6 19-20	5-8 ± 2.1	102 ± 7.5	15-11 ± 6.5	165 ± 12.0	28-4 ± 9.8

* One female without ovarian activity.   From Thimonier, Ravault, Ortavant, 1978.
G1: Normal variations of daylength at 47°N.
G2: 8 hrs of light
G3, G4, G5, G6: 8 hrs of light in 2 fractions : 7 hrs + 1 hr
The position of the 1-h light pulse relative to dawn is indicated.

From the measurement of prolactin levels in blood samples taken twice weekly, it was concluded that there is a seasonal variation in prolactin levels in ewes subjected to normal variation of daylength, the lowest being observed during the short daylengths and the highest during the long ones. This seasonal pattern even occurred under constant photoperiod. However, prolactin levels and amplitude of variation were the highest when the ewes received a light pulse 16 to 17 hours after dawn. Thus, there is a photosensitive phase during the diurnal cycle, identical to that observed in rams (Ravault and Ortavant, 1977).

B. With classical light treatments (Part I), sexual activity is promoted when there is a shift from long to short daylength.

In an experiment, Ile-de-France ewes were primed with a 1-hour light pulse given during the 17th hour after dawn (8 h of light in two fractions 7 h + 1 h) for more than 3 months from September 15 to beginning of January. The time of the 1-hour light pulse was then changed and, depending on the experimental group, was given either during the 8th, 11th, 14th or 19th hour after dawn or maintained during the 17th hour.

The following results were observed (Table 2) (Thimonier, 1981):

- the 17th hour-priming light pulse induced anoestrus 1 month earlier than in the controls,

- the change in the position of the light pulse induced an earlier subsequent onset of ovarian activity when the pulse was given during the 8th, 11th or 14th hour.

These treatments were able to advance the period of ovarian activity by about 5 months compared to that of the controls.

TABLE 2   Effect of a change in the position of the 1-h light pulse relative to dawn in ewes after priming with a light pulse at the 17th hour.

Priming period 15 september- 9 January	End of ovarian activity date ± s.d.	Position of the light pulse after the change (h after dawn)	Onset of ovarian activity	
			No of females with regular ovarian activity	Date ± s.d.**
7 h light +	18 December	7-8	7/8	[a] 28 Febr. ± 28.5 days
1 h 16 - 17 h	±	10-11	7/8	[a]2 March ± 11.3 days
after dawn	21.1 days	13-14	5/8	20 Febr. ± 8.6 days
		16-17	7/8	[b]9 May ± 11.7 days
		19-20	4/8	20 April ± 35.7 days
Control group*	17 January ± 11.6 days		7/7	[c]6 August ± 11.1 days

* Normal variations of daylength at 47°N.
** Values with different superscript letters are significantly different at $P < 0.05$.

From Thimonier, 1981 (J. Reprod. Fert.).

Measurement of prolactin levels in blood samples taken twice weekly clearly demonstrated a seasonal pattern in prolactin secretion, not related with the light treatment, the minimal values occurring at the same time as in the control ewes.

C. Earlier induction of anoestrus is also obtained (December 25 vs February 3 in ewes submitted to normal variations of daylength) when, from mid-October onwards, the ewes are submitted to a variable duration in daylength (normal variations in the twilight hours), i.e. a 1-hour light pulse is given 16 to 17 hours after a fixed dawn.

D. Finally, a period of ovarian is induced by beginning of March when, after priming with a light pulse 16 to 17 hours after dawn, the position of the 1-hour light pulse is changed to the 12th hour after dawn.

However, this induction is better (7 out of 8 ewes) when the twilight does not vary than when it follows the normal pattern (3 out of 8 ewes only).

PRACTICAL IMPLICATIONS AND CONCLUSIONS

From the preceding results, it is possible to conceive a photoperiodic treatment for the induction of oestrus and ovulation during anoestrus without having to use light-proof buildings, extra light being given only for short periods (Fig. 4).

Dawn is fixed by lights-on at a given hour in the morning and a light pulse of 1 hour is given 16 to 17 hours after dawn for 1.5 to 2 months. The time of the light pulse is then changed between the 10th and 12th hour, depending to the period of the year.

Ovarian activity is expected 1.5 to 2 months after the time of the light pulse has been changed. Other stimuli (ram effect, for example) can be associated, such stimuli being efficient for induction of oestrus and ovulation when the ewes are in shallow anoestrus due to the light.

Experiments in the field are necessary to demonstrate the efficiency of such management for the induction of fertile oestrus and ovulation during seasonal or post-partum anoestrus.

52

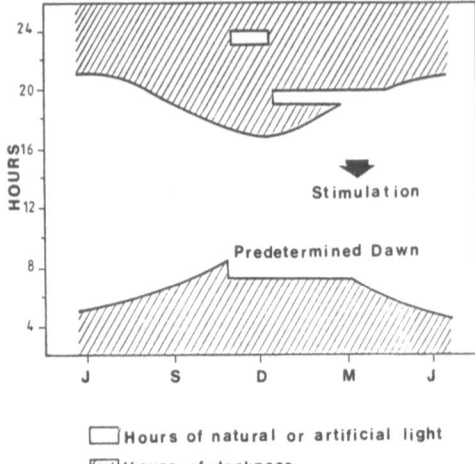

☐ Hours of natural or artificial light
▨ Hours of darkness

Fig. 4 A practical light treatment for induction of oestrus and ovulation
during anoestrus in the ewe.
A light pulse is first given 16 to 17 h after a fixed dawn then its
position is changed. At the expected period of the induced ovarian
activity, ewes are "stimulated" by teasing. Non pregnant ewes after a
first mating are expected to demonstrate cyclical oestrus.

REFERENCES

Bissonnette, T.H. 1941. Experimental modification of breeding cycles in
    goats. Physiol. Zool., 14, 379-381.
Clegg, M.T., Cole, H.H. and Ganong, W.F. 1965. The role of light in the
    regulation of cyclical estrous activity in sheep. U.S. Dep. agr. Mis.
    Pub., 1005, 95-103.
Dacheux, J.L., Pisselet, C., Blanc, M.R., Hochereau-de Reviers, M.T. and
    Courot, M. 1981. Seasonal variations in rete-testis fluid secretion
    and sperm production in different breeds of ram. J. Reprod. Fert.,
    61, 363-371.
Dutt, R.H. 1960. Temperature and light as factors in reproduction among
    farm animals. J. Dairy Sci., 43, Suppl., 123-144.
Evans, G. and Robinson, T.J. 1980. Reproductive potential and
    endocrinological responses of sheep kept under controlled lighting.
    I. Comparative reproductive performance of four breed types of ewe.
    Anim. Reprod. Sci., 3, 23-37.
Godley, W.C., Wilson, R.L. and Hurst, V. 1966. Effect of controlled
    environment on the reproductive performance of ewes. J. Anim. Sci.,
    25, 212-216.
Hackett, A.J. 1982. Effect of dose of pregnant mares' serum gonadotrophin
    on the reproductive performance of ewes synchronized for estrus and
    housed in total confinement. Can. J. Anim. Sci., 62, 291-294.
Heaney, D.P., Ainsworth, L., Batra, T.R., Fiser, P.S., Hackett, A.J.,
    Langford, G.A, and Lee, A.J. 1980. Recherches pour la production
    intensive du mouton en stabulation. Agric. Can. Anim. Res. Inst.
    Tech. Bull., n° 2, 57 pp.
Karsch, F.J., Bittman, E.L., Foster, D.L., Goodman, R.L., Legan, S.J. and
    Robinson, J.E. 1984. Neuroendocrine basis of seasonal reproduction.
    Recent Prog. Horm. Res., 40, 185-232.

Land, R.B., Pelletier, J., Thimonier, J. and Mauléon, P. 1973. A quantitative study of genetic differences in the incidence of oestrus, ovulation and plasma luteinizing hormone concentration in the sheep. J. Endocr., 58, 305-317.

Legan, S.J. and Karsch, F.J. 1980. Photoperiodic control of seasonal breeding in ewes: modulation of the negative feedback action of estradiol. Biol. Reprod., 23, 1061-1068.

Legan, S.J. and Karsch, F.J. 1983. Importance of retinal photoreceptors to the photoperiodic control of seasonal breeding in the ewe. Biol. Reprod., 29, 316-325.

Mauléon, P. and Rougeot, J. 1962. Régulation des saisons sexuelles chez des brebis de races différentes au moyen de divers rythmes lumineux. Ann. Biol. anim. Bioch. Biophys., 2, 209-222.

Ortavant, R. 1977. Photoperiodic regulation of reproduction in the sheep. In: "Management of reproduction in sheep and goats". Symposium, Madison, 24-25 July, Univ. Wisconsin, pp 58-71.

Pelletier, J. 1971. Influence du photopériodisme et des androgènes sur la synthèse et la libération de LH chez le bélier. Thèse Doct. Etat ès Sci. Nat., Univ. Paris, 243 pp.

Pelletier, J., Blanc, M., Daveau, A., Garnier, D.H., de Reviers, M.M. and Terqui, M. 1981. Mechanism of light action in the ram: a photosensitive phase for LH, FSH, testosterone and testis weight? In: "Photoperiodism and reproduction in vertebrates", (Eds R. ORTAVANT, J. PELLETIER and J.P. RAVAULT), (INRA, Paris), pp. 117-134.

Ravault, J.P. and Ortavant, R. 1977. Light control of prolactin secretion in sheep. Evidence for a photoinducible phase during a diurnal rhythm. Ann. Biol. anim. Bioch. Biophys., 17, 459-473.

Ravault, J.P., Daveau, A. and Ortavant, R. 1981. Evidence for a photosensitive phase for prolactin secretion in relation to the dusk in rams. In: "Photoperiodism and reproduction in vertebrates". (Eds R. ORTAVANT, J. PELLETIER and J.P. RAVAULT), (INRA, Paris), pp. 135-146.

Robinson, J.J., Fraser, C. and McHattie, I. 1975. The use of progestagens and photoperiodism in improving the reproductive rate of the ewe. Ann. Biol. anim. Bioch. Biophys., 15, 345-352.

Rougeot, J. 1969. Accélération du rythme de la reproduction chez le mouflon de Corse (Ovis ammon musimon Schreber, 1782) au moyen de cycles photopériodiques semestriels. Ann. Biol. anim. Bioch. Biophys., 9, 441-443.

Rouger, Y., 1974. Etude des intéractions de l'environnement et des hormones sexuelles dans la régulation du comportement sexuel des Bovidae. Thèse Doct. Etat ès Sci. Nat., Univ. Rennes.

Terqui, M., Delouis, C. and Ortavant, R. 1984. Photoperiodism and hormones in sheep and goats. In: "Manipulation of growth in farm animals" (Eds J.F. ROCHE and D. O'CALLAGHAN), Martinus Nijhoff, The Hague, pp. 246-259.

Thimonier, J. and Mauléon, P. 1969. Variations saisonnières du comportement d'oestrus et des activités ovarienne et hypophysaire chez les ovins. Ann. Biol. anim. Bioch. Biophys., 9, 233-250.

Thimonier, J. 1975. Etude de la puberté et de la saison sexuelle chez les races prolifiques et leurs croisements avec des races françaises. In: "Les Races Prolifiques". 1ères Journées de la Recherche Ovine et Caprine, Paris 2-3-4 décembre, (Eds INRA-ITOVIC-SPEOC), II, 18-37.

Thimonier, J., Ravault, J.P. and Ortavant, R., 1978. Plasma prolactin variations and cyclic ovarian activity in ewes submitted to different light regimens. Ann. Biol. anim. Bioch. Biophys., 18, 1229-1235.

Thimonier, J. 1981. Control of seasonal reproduction in sheep and goats by light and hormones. J. Reprod. Fert., Suppl. 30, 33-45.

Thimonier, J. and Gauthier, D. 1984. Seasonality of reproduction in cattle and sheep and its consequences on reproduction management. In: "The reproductive potential of cattle and sheep", (Eds. R. ORTAVANT and H. SCHINDLER), INRA Paris, pp 133-153, .

Thwaites, C.J. 1965. Photoperiodic control of breeding activity in the Southdown ewe with particular reference to the effects of an equatorial light regime. J. agric. Sci., Camb., 65, 57-64.

Vesely, J.A. 1975. Induction of lambing every eight months in two breeds of sheep by light control with or without hormonal treatment. Anim. Prod., 21, 165-174.

Walrave, Y., Cantin, P., Desvignes, A. and Thimonier, J. 1975. Variations saisonnières de l'activité sexuelle des races ovines du Massif Central. In: "Les Races Prolifiques". 1ères journées de la Recherche Ovine et Caprine, Paris, 2-3-4 décembre, (Eds. INRA-ITOVIC-SPEOC), II, 261-271.

Wodzicka-Tomaszewska, M., Hutchinson, J.C.D. and Bennett, J.W. 1967. Control of the annual rhythm of breeding in ewes: effect of an equatorial daylength with reversed thermal seasons. J. agric. Sci., Camb., 68, 61-67.

Yeates, N.T.M. 1949. The breeding season of the sheep with particular reference to its modification by artificial means using light. J. agric. Sci., Camb., 39, 1-43.

EFFECTS OF MELATONIN ON THE TIME OF ONSET OF THE BREEDING SEASON IN DIFFERENT BREEDS OF SHEEP

J.F. Roche*, J.P. Hanrahan**, J.F. Quirke***
E. Ronayne*

*Faculty of Veterinary Medicine,
Ballsbridge, Dublin 4, Ireland
**The Agricultural Institute,
Belclare, Tuam, Co. Galway.
***The Agricultural Institute,
Dunsany, Co. Meath,Ireland.

ABSTRACT

An experiment was carried out to determine if oral administration of 3 mg of melatonin given at 16.00 h each day, thrice weekly or once weekly would advance the breeding season in ewes compared to ewes implanted with melatonin or control ewes. The breeding season was advanced by daily oral administration or by use of implants of metatonin. In a second experiment, the efficacy of melatonin implants to advance the breeding season in three breeds was studied. Melatonin implants given in June/July significantly advanced the breeding season in Suffolk, Texel and Scottish Blackface ewes. These results indicate a role for the pineal gland through its secretion of melatonin as an important mediator of the effects of photoperiod on seasonality of breeding in ewes. In addition, the results show that continuous administration of melatonin by implant was as effective as daily oral administration at 16.00 h, thus obviating the need for administration of melatonin at a specific time each day.

INTRODUCTION

The ewe is a seasonally breeding animal with oestrus and ovulation occurring on a recurrent 16-17 day basis in non-pregnant animals from September to March in northern latitudes. Rams also have a similar annual cycle of sexual activity with both testes size and sperm output at a maximum in the autumn (Lincoln, 1979). The onset of cyclicity is achieved through a complex interaction between environmental signals, and the interpretation of these signals by certain brain centres which result in seasonal changes in hypothalamic and hypophyseal function. There is clear evidence that day length is a major environmental factor influencing seasonality of breeding in ewes. Exposure of anoestrous ewes to short daylength will induce premature oestrous cycles (Marshall, 1937; Ortavant et al.,1964; Yeates, 1949).

It is well known that the pineal gland is an important mediator of the effects of daylength on seasonality of breeding in small animals such as ferrets, hamsters and voles. However, in farm animals, the evidence is less clear cut. Superior cervical ganglionectomy in mares (Sharp et al., 1979) and rams (Lincoln, 1979) can alter seasonal reproductive patterns but initial studies with pinealectomized ewes (Roche et al., 1970) failed to show that the pineal gland was an important mediator of the effects of daylength on seasonality of breeding in ewes. It is possible that when the photoperiodic signal is removed in pinealectomized ewes, the animals simply use some other environmental cue to differentiate between seasons of the year. Alternatively, studies over one breeding season may have given insufficient time for the effects of pinealectomy to manifest themselves on seasonality of breeding. However, in long term studies, Seamark et al. (1981) also showed that that pinealectomy had little effect on the onset of breeding activity in ewes over a two year period.

More recently, Bittman et al. (1983) exposed pinealectomized ewes to abrupt changes in photoperiod in light controlled rooms, and were able to demonstrate clearly that the pineal gland is required for induction of oestrous cycles in ewes challenged with a short day photoperiod.

Thus, there is increasing evidence to show that the pineal gland is an important mediator of daylength effects in the ewe, but obviously the ewe can equally use cues other than daylength in order to maintain seasonality of cyclicity. The identification of the active photoperiodic hormone from the pineal gland has been helped by work from laboratory species suggesting that melatonin, an indoleamine which is synthesized and released by the pineal, is the

putative mediator regulating neuroendocrine-gonadal function
(Minneman and Wurtman, 1975).

Exogenous melatonin can mediate photoperiodic effects
in hamsters (Turek et al., 1975) and ferrets (Thorpe et al.,
1974). More recently, Rollag, O`Callaghan and Niswender
(1978), showed that peripheral concentrations of melatonin
accurately reflect the duration of darkness in ewes, and
these authors hypothesized that light inhibits the secretion
of melatonin. Thus, melatonin may be the hormone mediating
the effects of photoperiod on reproduction in the ewe. Data
from Bittman et al. (1983) are in agreement with this
hypothesis, because pinealectomy obliterated the night time
rise in melatonin in ewes. In summary, the pineal gland
through its light controlled secretion of melatonin, appears
to be an important organ in the regulation of seasonal
breeding in ewes.

INDUCTION OF OESTRUS IN ANOESTROUS EWES
The restricted breeding season of the ewe is an
important constraint to reproductive efficiency on an annual
basis. This can be overcome to some extent by hormone t-
herapy based on the use of progestagen impregnated pessaries
and pregnant mares serum gonadotrophin (PMSG) injections
(Gordon, 1975; Cognie and Mauleon, 1983). The response to
progestagen-PMSG treatment, however, in terms of both
conception rate and litter size obtained, depends on the
time of year the treatment is initiated. Best results are
obtained during the late anoestrous period. However, when
this hormone treatment is applied during the early
anoestrous period only a small proportion of the ewes which
fail to conceive at the induced oestrus will continue to
ovulate, with the majority lapsing into anoestrus. Thus, it
is important to attempt to develop methods which will
advance the breeding season and allow all ewes to continue
to cycle.

The manipulation of the photoperiod to which ewes are

58

exposed is one way to achieve continued cyclicity but this
is not practical for routine lamb production. An
alternative approach is to expose ewes to physiological
short days under normal environmental conditions by giving
them melatonin from 16.00 h onwards during the summer, since
melatonin appears to be the mediator of the effects of
photoperiod on reproductive function in the ewe. In fact
Nett and Niswender (1982), Kennaway, Gilmore et al. (1982)
and Arendt et al. (1983) have successfully advanced the
breeding season in ewes by this method. The aim of the
research reported here was to determine if implants of
melatonin would be as effective as once a day or three times
weekly administration of melatonin given orally.

MATERIALS AND METHODS

     Melatonin implants were made from silastic sheeting
(Dow Corning Ltd., U.S.A.,500-1 and .005 ins. thick).
Pieces of silastic 7 cm x 5 cm were prepared and medical
grade adhesive spread around the ends of one half. Then 700
mg melatonin (Sigma Ltd., U.K.) were placed on the sheeting,
which was folded over and the ends sealed. The implants
were placed subcutaneously in the axillary region. In the
animals which received melatonin orally, the compound was
first dissolved in methanol and diluted with distilled water
so that 30 ml contained 3 mg melatonin. The 3 mg melatonin
in 30 ml was administered using an oral drenching gun.

Experiment 1: Fifty four Cheviot type mature ewes and ten
Suffolk x Cheviot ewe lambs were used. The ewes were
randomized on a weight basis to the following treatments:
     (i)   Control group - dosed daily with
              physiological saline
     (ii)   Melatonin implants - given as described
              above
     (iii)   Dosed daily with 3 mg melatonin at 16.00 h
     (iv)   Dosed on Monday, Wednesday and Friday as in
              at 16.00 h
     (v)   Dosed once weekly as in (iii)

The ewe lambs were randomized on a weight basis to
treatments (i) and (iii). Treatment started in the fourth
week of July and continued until the 7th of October. The
ovaries of all ewes were examined by endoscopy on 28th
September. Blood samples were taken from all animals on
Monday, Wednesday and Friday each week beginning four weeks
after start of melatonin oral administration. These samples
were analyzed for progesterone by an enzyme immunoassay
system previously validated (Cleere et al., 1984).

Experiment 2: In this trial thirty seven Suffolk, thirty
three Texel and twenty one Scottish Blackface mature ewes
were allocated within breed to either a control group or to
a treatment group where ewes received the melatonin implants
described earlier. The Suffolk and Texel ewes were implanted
on June 21st and the Scottish Blackface ewes received their
implants on July 4. Vasectomised rams were joined on the day
of implantation and the ewes were inspected daily for raddle
marks until first oestrus was detected. Ewes were
laparascoped on three occassions following implantation (cf
Tables 2 and 3) to determine if ovulation had occurred. In
the case of Suffolk and Texel ewes fertile rams were joined
following the commencement of cyclicity and the implants
were removed subsequently.

RESULTS
    Experiment 1: The data presented in Table 1 indicate
clearly that daily oral administration of melatonin was
effective in advancing the time of first ovulation and the
time of detection of the first significant rise in
progesterone in peripheral blood. However, giving melatonin
orally thrice weekly or once weekly had no effect on the
time of first ovulation. The ewes that received the
implants of melatonin also had an earlier onset of first
ovulation and first rise in progesterone in blood. In the
case of the ewe lambs, oral adminsitration of melatonin also
advanced the time to first ovulation.

60

Table 1   The number of ewes which had ovulated at
          laparoscopy and with elevated levels of
          progesterone (>0.80 ng/ml)in peripheral plasma
          before September 28.

		No. of Ewes	
	No.of Ewes	Ovulated	With high Progesterone
ADULT EWES:			
Control	12	5	7
Oral melatonin daily	12	12	11
Oral melatonin thrice weekly	11	6	3
Oral melatonin once weekly	10	3	3
Melatonin implants	9	8	7
EWE LAMBS:			
Control	5	0	0
Oral melatonin daily	5	5	2

Experiment 2:

One Suffolk, five Texel and six Scottish Blackface ewes lost
their melatonin implants and these animals have been
excluded from the results (Tables 2 and 3). In the Suffolk
ewes, the mean advance in date of first oestrus was 7.5+ 3.3
days (P<0.05) and the number of days from implantation of
melatonin to first oestrus was 58+1.4 days (Table 2).

   The mean advance in days to first oestrus for the Texel
ewes was 10.5+3.9 days (P<0.01) and the number of days from
insertion of implants to first oestrus was 73.2+2.2.   In

the Suffolk and Texel ewes which carried melatonin implants there was no treatment effect on subsequent pregnancy rate.

TABLE 2: Effect of implants of melatonin on onset of the breeding season in Scottish Blackface ewes.

	Control	Implant
No. of Ewes	10	5
No. of ewes which had ovulated by:		
Sept. 2	1	5
Sept. 19	3	5
Oct. 7	8	5
Date of first oestrus		
- Median	Oct.17	Sept. 19
- Mean	Oct.17$\pm$5.6	Sept.19$\pm$1.6
Mean advance in date of first oestrus (days)	28.5$\pm$6.7***	
Days from implantation to oestrus		77$\pm$1.6

*** P<0.01

The data presented in Table 2 show that in the Scottish Blackface ewes, the mean date to first oestrus was advanced by 28.5$\pm$6.7 days (P< 0.01) and the number of days from inserion of implants to oestrus in the implanted ewes was 77$\pm$1.6 days.

DISSCUSSION

The effect of melatonin treatment in hamsters is dependent on the photoperiod to which the animals are exposed and it appears that the role of melatonin is not to directly alter hypothalamic-pituitary function, but to affect brain centres responsible for interpreting the photoperiod. There is a complex relationship between the action of melatonin and the photoperiodic cues received

TABLE 3: Effect of implants of melatonin on onset of the breeding season in suffolk and Texel ewes

| | Breed | | | |
| | Suffolk | | Texel | |
	Control	Implant	Control	Implant
No. of ewes	17	19	15	13
No. ewes which had ovulated by:				
- July 20th	1	1	0	0
- Aug. 9th	8	19	2	0
- Aug 18th	-	-	8	13
Date of first oestrus				
- Median	Aug. 25	Aug. 17	Sept. 6	Aug. 28
- Mean	Aug. 26 ± 2.9	Aug. 19 ± 1.4	Sept. 13 ± 3.2	Sept. 2 ± 2.2
Mean advance in date of first oestrus (days)		$7.5\pm3.3**$		$10.5\pm3.9***$
Days from implant to first oestrus		58.8 ± 1.4		73.2 ± 2.2

$**P<0.05$ $***P<0.01$

(Losee and Turek, 1981). However, in the ewe exposed to
decreasing daylength, administration of melatonin orally is
a reliable method to maintain blood concentrations of
melatonin over an 8 hr period above normal day time
concentrations (Nett and Niswender, 1982; Kennaway et al.,
1982). In addition, such administration can advance the
breeding season in ewes. The results presented here confirm
that oral administration of melatonin at 16.00 h does
significantly advance the onset of the breeding season in
different ewe breeds.

If the photosensitive pineal gland through its
secretion of melatonin modulates the effects of changing
photoperiod on seasonality of breeding in ewes, then
continuous administration of melatonin by an implant should
be photoperiodically similar to constant darkness. This
raises the question whether or not continuous administration
of melatonin would be interpreted as a stimulatory
photoperiodic cue. The data presented from the present
experiments clearly show that continuous administration of
melatonin, when ewes are on a naturally decreasing
daylength, is interpreted as stimulatory. However, it is
not clear what the endogenous changes in concentrations of
melatonin throughout a twenty four hour period are in the
ewes receiving melatonin implants in these experiments since
these determinations have not yet been performed. In
addition, it is not clear how melatonin stimulates the onset
of the breeding season. Further research is required but
these data do add to the fact that the pineal gland, through
its secretion of melatonin, is an important mediator of the
effects of photoperiod on seasonality of reproduction in the
ewe.

ACKNOWLEDGEMENTS

The authors wish to acknowledge the help of Mr. D.
O'Callaghan, Mr. D. Prendiville and Mr. P. Madden in the
conduct of these experiments. The technical help of Ms. A.

Whelan and Ms. S. White in carrying out the progesterone
assays is also acknowledged.

REFERENCES

Arendt, J., Symons, A.M., Laud, C.A. and Pryde, S.J. 1983.
        Melatonin can induce early onset of the breeding
        season in ewes. J. Endocrinology, 97, 395-400.
Bittman, E.L., Karsch, F.J. and Hopkins, 1983. Role fo the
        pineal gland in ovine photoperiodism: Regulation of
        seasonal breeding and negative feedback effects of
        estradiol upon luteinizing hormone secretion.
        Endocrinology, 113, 329-336.
Cleere, W., Gosling, J.P., Morris, N.V., Charleton, N.S.,
        Moloney,B.T., Sreenan, J.M. and Fottrell, P.F.
        1984. A high performance, high throughput enzyme
        immunoassay for the analysis of progesterone in
        plasma or milk. Irish Vet. J.(In press).
Cognie, Y. and Mauleon, P. 1983. Control of reproduction in
        the ewe. In: Sheep Production. Edited by W.
        Haresign, Chapter 18, pp 381-392. Butterworths,
        London.
Gordon, I. 1975. Hormonal control of reproduction in sheep.
        Proc. Br. Soc. Anim. Prod., 4 (New Series), 79-93.
Kennaway, D.J., Gilmore, T.A. and Seamark, R.F. 1982.
        Effect of melatonin feeding on serum prolactin and
        gonadotrophin levels and the onset of seasonal
        oestrous cyclicity in sheep. Endocrinology, 110,
        1766-1772.
Lincoln, G.A. 1979. Photoperiodic control of seasonal
        breeding in the ram: participation of the cranial
        sympathetic nervous system. J. Endocrinology, 82,
        135-147.
Losee, S.H. and Turek, F.W. 1981. In: Pineal Function.
        Edited by C.D. Matthews and R.F. Seamark; pp.
        67-75. Elsevier/North Holland Biomedical Press.
Marshall, F.H.A. 1937. On the change-over in the oestrous
        cycle in animals after transference across the
        equator with further observations on the incidence
        of the breeding seasons and the factors controlling
        sexual periodicity. Proc. Royal Society Series B,
        122, 413-428.
Minneman, K.P. and Wurtman, R.J. 1975. Effects of pineal
        compounds of mammals. Life Science, 17, 1189-1200.
Nett, T.M. and Niswender, G.D. 1982. Influence of exogenous
        melatonin on seasonality of reproductin sheep.
        Theriogenology, 17, 645-653.
Ortavant, R., Mauleon, P. and Thibault, C. 1964.
        Photoperiodic control of gonadal and hypophyseal
        activity in domestic animals. Annals of the New
        York Academy of Sciences, 11, 157-193.
Roche, J.F., Karsch, F.J., Foster, D.L., Takagi, S. and
        Dziuk, P.J. 1970, Effect of pinealectomy on estrus,

ovulation and luteinizing hormone in ewes.
Biology of Reproduction, 2, 251-254.

Rollag, M.D., O'Callaghan, P.L. and Niswender, G.D. 1978.
Serum melatonin concentrations during different
stages of the annual reproductive cycle in ewes.
Biology of Reproduction, 18, 279-285.

Seamark, R.F., Kennaway, D.J., Matthews, C.D., Fellenberg,
A.J., Phillipou, G., Kotaras, P., McIntosh, J.E.A.,
Dunstan, E. and Obst, J.M. 1981. The role of the
pineal gland in seasonality. J. Reprod. Fert.,
Suppl. 30, 15-21.

Sharp, D.C., Vernon, M.W. and Zavy, M.J. 1979. Alterations
of seasonal reproductive patterns in mares
following superior cervical ganglionectomy. J.
Reprod. Fert., Suppl. 27, 87-93.

Thorpe, P.A. and Herbert, J. 1974. The role of the pineal
gland in the response of ferrets to artificially
restricted photoperiods. J. Endocrinology, 63,
56p.

Turek, F.W., Desjardins, C. and Menaker, M. 1975. Melatonin:
Antigonadal and progonadal effects in male golden
hamsters. Science, 190, 280-282.

Yeates, N.T.M. 1949. The breeding season of the sheep with
particular reference to its modification by means
of artificial light. J. Agric. Sci. Camb., 39,
1-49.

66

INDUCTION OF OVULATION IN ANOESTROUS EWES USING GONADOTROPHINS

A.S. McNeilly, J.M. Wallace and D.T. Baird

MRC Unit of Reproductive Biology and Department of Obstetrics and
Gynaecology, University of Edinburgh, Centre for Reproductive Biology,
37 Chalmers Street, Edinburgh EH3 9EW, Scotland

ABSTRACT
    Seasonal anoestrus is associated with a reduction in the frequency of
pulsatile release of LH in all breeds of sheep examined so far. We have
investigated the gonadotrophic requirements for ovulation and formation of
a corpus luteum in 4 breeds of sheep with differing lengths of anoestrus
by infusing LH in a pulsatile manner to reproduce the pattern observed
during the follicular phase of the oestrous cycle with or without addition
of FSH (a) In Finn-Merino ewes with "shallow" anoestrus, pulsatile LH
administration alone induces ovulation and normal luteal function. (b) In
Scottish Blackface ewes with a long "deep" anoestrus, neither pulsatile LH
nor FSH induced ovulation. (c) In Welsh and Damline ewes with a moderate
degree of anoestrus, pulsatile LH induced ovulation but with a high
incidence of inadequate corporus luteum (C.L.) function. Addition of FSH
in the dose and duration used did not significantly increase the incidence
of normal C.L. It is concluded that the gonadotrophic requirement for
follicular development, ovulation and formation of a normal C.L. during
seasonal anoestrus vary depending on the genetic background of the sheep.

    INTRODUCTION

    Seasonal anoestrus in the ewe is associated with a reduced frequency

of spontaneous pulsatile discharges of LH compared to the breeding season

(Scarammuzzi and Baird, 1977; Yuthasastrakosol et al., 1977). Since

plasma levels of FSH appeared to be normal (McNeilly and Land, 1979;

McNeilly et al., 1982; McNatty et al., 1984; Walton et al., 1977), we

suggested that the failure of normal follicular development during

anoestrus was related solely to the low frequency of LH pulses. To test

this hypothesis we initially investigated the effectiveness, in inducing

ovulation of injecting ovine LH in amounts sufficient to simulate

endogenous LH pulses at a frequency changing in a manner equivalent to the

normal pattern of pulsatile LH release in the periovular period in the ewe

(Baird, 1978). Having successfully induced ovulation with normal luteal

function in Finn-Merino ewes in mid anoestrus (McNeilly et al., 1982) we

have extended the investigation to study the effectiveness of the same

regime for the injection of LH in Scottish Blackface, Welsh and Damline

ewes. In addition, as a result of the experiment with Scottish Blackface

ewes we have investigated the effect of concomittant treatment with FSH on

the response to LH in Welsh and Damline ewes, the result of which will be

published in detail elsewhere. (Wallace et al., 1984) This paper

summarized and compares our results in the four breeds of ewes with

differing lengths and depth of anoestrus.

MATERIALS AND METHODS

Animals and experimental design

All experiments were carried out at Dryden Field Station, Roslin Midlothian, Scotland between June and August. Animals were kept in individual pens or in metabolism crates and had access to food and water ad libitum. The dates, numbers of animals per group and treatments are summarized in Table 1.

TABLE 1   Details of breeds of ewe and experiments carried out in seasonal anoestrus.

	Duration of anoestrus	Time of Expt.	No per Treatment	Treatments			
				LH	FSH	LH + FSH	Saline
Finn-Merino	May-Sept	Aug '79	4	+	−	−	−
Scottish Blackface	Feb-Oct	July '81	7	+	+	−	+
Welsh	Mar-Oct	June '83	5	+	+	+	+
Damline*	Apr-Sept	June '83	5	+	+	+	+

* Damline ewes are a cross as follows: 47% Finnish Landrace, 24% East Friesland, 17% Border Leicester and 12% Dorset Horn.

In all experiments, LH was injected for 72h with an increasing frequency, once every 3h for 24h, once every 2h for 24h and once every hour for 24h. For LH injections a stock of ovine LH (NIH-LH-S13 for experiments in Finn- Merino and Scottish Blackface ewes, NIH-LH-S23 for Welsh and Damline ewes) was prepared in 0.9% sodium chloride solution and diluted in saline containing 10% sheep plasma. Each injection of LH (10ug), an amount which resulted in an LH pulse similar in magnitude and duration to that seen in the normal preovulatory period (McNeilly, et al 1982) was given by syringe (1-2 mg over 10-20s) via a jugular cannula.

Ovine FSH was prepared in the same way as LH. In Scottish Blackface ewes, FSH (NIH-FSH-S6) was injected at the same frequency as LH but at a decreasing dose per injection i.e 500 ug/3h for 24h, 200 ug/2h and 50 ug/h for 24h. In Welsh and Damline ewes, FSH (NIH-FSH-S14) was injected at a concentration of 30 ug in 3mls every 3h for 24h and then at 20ug in 2mls every 2h for 12h.

Assessment of response to treatment

In all experiments blood samples were taken at regular intervals
(30 min - 3 hourly) throughout the period of treatment to assess the
changes in plasma levels of LH, FSH and prolactin. Following treatment
blood samples were taken daily for measurement of progesterone as an
assessment of corpus luteum function. The initial studies which have been
reported in detail previously (McNeilly et al., 1982) were carried out in
4 Finnish Landrace x Merino ewes with ovarian transplants and 2 with
utero-ovarian transplants. Using these ewes it was possible to monitor
the change in ovarian steroid secretion in response to the treatment with
LH. However, no assessment of ovulation rate could be made but, as in the
other experiments, the adequacy of luteal function was assessed by
measurement of plasma progesterone.

Scottish Blackface ewes were sent to slaughter 6 days after the end of
treatment and ovulation and follicular development was assessed by
dissection of the ovaries.

In both Welsh and Damline ewes, ovulation was assessed by
laparoscopic examination of the ovaries 6 or 7 days after the final
hormone injection.

All hormones were measured by specific radioimmunoassays as reported
in detail previously (see McNeilly et al., 1982 for references to
methods).

RESULTS
Finn-Merino ewes

Since these results have been reported previously (McNeilly et al.,
1982) they will only be summarized. The pulsatile injection of LH at
increasing frequency resulted in an increase in ovarian secretion of
oestradiol, a preovulatory surge of both LH and FSH and normal corpus
luteum function. The pattern of secretion of LH, FSH, 17B oestradiol and
progesterone was similar to that occuring in the oestrous cycle in these
ewes, with the preovulatory surge occuring at a similar time (52-57h)
after the first injection of LH as after the injection of prostaglandin to
induce luteal regression in the normal oestrous cycle (52-68h).

Scottish Blackface ewes

In contrast to the results obtained in Finn-Merino ewes, injection of
LH alone failed to induce ovulation as assessed by dissection of the ovary
(Table 2). Indeed there was no increase in plasma levels of progesterone

and no evidence of corpora albicans within the ovary which might have suggested ovulation of a follicle but subsequent failure to maintain the resulting corpus luteum.

TABLE 2   Effect of injection of LH or FSH over a 72h period on ovarian weight, maximum follicle size and ovulation in anoestrous Scottish Blackface ewes. Results are mean $\pm$ s.e.m.

Treatment	n	Ovarian wt (g)	Max follicle diameter (mm)	Ovulation rate
Saline	7	1.56 $\pm$ 0.35	4.85 $\pm$ 1.22	0
LH	7	1.71 $\pm$ 0.36	6.00 $\pm$ 1.41	0
FSH	7	2.07 $\pm$ 1.29	4.14 $\pm$ 1.87	0

Treatment with LH or FSH also had no significant effect on ovarian weight or size of the largest follicle (Table 2). All ewes receiving LH showed a significant increase in LH between 37 – 49h after first injection of LH but the amplitude of this increase (23 ± 5 ng/ml) was substantially below the normal preovulatory surge of LH seen in the normal oestrous cycle (85 ± 6 ng/ml).

Welsh and Damline ewes

The experiments in Welsh and Damline ewes were carried out at the same time in order to provide a direct breed comparison of the effectiveness of treatment with pulses of LH alone or in combination with FSH. Thus the results will be reported together. During laparoscopic assessment of ovulation after treatment, the presence of corpora lutea and or corpora albicans were noted in ewes previously treated with LH or LH and FSH. Since no corpora lutea or corpora albicans were seen in any of the ewes treated with saline or FSH alone, ovulation rate has been assessed by combining both corpora lutea and corpora albicans. The incidence of ovulation and normal corpus luteum function is summarized in Table 3.

Treatment with LH alone induced ovulation, assessed as corpora lutea plus corpora albicans, in all Welsh and Damline ewes. However, only one of the 5 Damline ewes had a corpus luteum although this was normal in terms of progesterone secretion. In contrast 4 of the 5 Welsh ewes had visually normal corpora lutea although only 2 or these were functionally normal in terms of progesterone. Thus in both breeds, although the treatment with LH alone induced ovulation, the incidence of inadequate

TABLE 3   Presence of corpora lutea (CL), corpora albicans (CA) normality
           of CL function in terms of progesterone secretion (normal) in
           anoestrus Welsh and Damline ewes induced to ovulate with LH or
           LH and FSH.

TREATMENT	WELSH			DAMLINE		
	CL	CL + CA	NORMAL	CL	CL + CA	NORMAL
LH	4/5	5/5	2/5	1/5	5/5	1/5
LH + FSH	3/5	4/5	1/5	1/5	4/4	3/4

When FSH was given concomitantly with LH, a similar proportion of
Welsh ewes had corpora lutea but most were non functional. In contrast
all Damline ewes now had corpora lutea, 3 of the 4 showing normal luteal
levels of progesterone.

The ovulation rates in the two breeds did not differ significantly
although they tended to be higher than in the breeding season in Welsh
ewes, but similar in the Damline ewes (Table 4).

TABLE 4   Parameters of the preovulatory LH surge and ovulation rate in
           anoestrus ewes induced to ovulate with LH or LH and FSH:
           comparison with the breeding season.

Parameter		Anoestrus	
	Breeding Season	LH	LH + FSH
Damline ewes			
Time to onset of LH (h)	55 ± 4	29 ± 5	26 ± 1
Max LH peak ht (ng/ml)	91 ± 9	40 ± 4	37 ± 8
Ovulation rate	2.3 ± 0.3	1.6 ± 0.3	2.5 ± 0.6
(range)	(1-3)	(1-2)	(1-4)
Welsh ewes			
Time to onset of LH (h)	50 ± 8	54 ± 7	25 ± 2
Max LH peak ht (ng/ml)	137 ± 20	55 ± 17	72 ± 11
Ovulation Rate	1.3 ± 0.1	1.8 ± 0.5	1.8 ± 0.8
(Range)	(1-2)	(1-3)	(1-4)

In both Welsh and Damline ewes, the preovulatory surge of LH was significantly smaller than in the breeding season, and except in Welsh ewes treated with LH alone, the surge occurred significantly earlier than in the breeding season if one compares time from the first injection of LH in anoestrus to the time from injection of prostaglandin to induce luteal regression in the breeding season (Table 4).

Comparison of pulsatile secretion of LH prior to treatment showed similar frequencies in both breeds in anoestrus but these were significantly less than during the luteal phase in the breeding season. Amplitude of each pulse was significantly greater in anoestrus than the breeding season in both breeds. (Table 5)

TABLE 5   Pulsatile secretion of LH in anoestrus compared to the breeding season.

Breed	Season	Basal (ng/ml)	Frequency (per 12h)	Amplitude (ng/ml)
Welsh	Breeding	0.9 ± 0.2	3.6 ± 0.6	3.6 ± 0.6
	Anoestrus	1.1 ± 0.1	2.3 ± 0.5	10.1 ± 2.3
Damline	Breeding	1.3 ± 0.2	3.6 ± 0.4	2.9 ± 0.6
	Anoestrus	1.1 ± 0.3	2.5 ± 0.5	8.1 ± 1.5

In contrast plasma levels of FSH, although tending to be lower, were not significantly different in anoestrus compared to the luteal phase of the breeding season (Table 6).

DISCUSSION

The studies summarized here show that injection of LH alone at an increasing frequency to mimic the natural pattern in the preovulatory period in the ewe will induce a preovulatory LH surge and ovulation in Finn-Merino, Welsh and Damline ewes, but not in Scottish Blackface ewes. In terms of ovulation rate and "depth" of anoestrus, it might be assumed that Welsh and Scottish Blackface ewes would have comparable results.   The reason for the failure of Scottish Blackface ewes to ovulate remains unknown. It may relate to the fact that the experiments in Scottish Blackface ewes were carried out in mid July towards the latter end of mid anoestrus while experiments in Welsh ewes were carried out at a relatively earlier stage of anoestrus. In Ile de France ewes it has been shown that the ability of LH or LH and FSH injections to induce ovulation is dependent on the stage of anoestrus, treatments being more effective in early anoestrus (Oussaid, 1982).

TABLE 6 Plasma levels of FSH in Breeding season and anoestrus.

BREED	FSH (ng/ml)	
	BREEDING SEASON	ANOESTRUS
Finn–Merino	90 + 9	88 + 4
Scottish Blackface	88 + 10	65 + 5
Welsh	78 + 7	61 + 7
Damline	56 + 5	43 + 5

The pulsatile secretion of LH in the present study is more frequent than the one pulse per 12h reported previously for Welsh ewes in anoestrus (Martensz et al., 1979). In the Scottish Blackface ewes, pulsatile secretion was certainly less than the Welsh ewes in these experiments.

Again, in terms of "depth" of anoestrus and the fact that both breeds of ewe are 50% Finnish–Landrace, it might be expected that Finn–Merino ewes would respond in a similar manner to Damline ewes. As with Welsh and Scottish Blackface ewes, this is not the case since LH injections alone induced ovulation with normal corpus luteum function in 4 of 4 Finn–merino ewes but only one of 5 Damline ewes formed a normal corpus luteum.

However this is difficult to reconcile with the fact that FSH levels in both Damline and Welsh ewes do not alter within breeds between the breeding season and anoestrus (Table 6). Thus, if the failure of LH alone to induce ovulation with normal luteal function in Damline ewes was due to lack of FSH, how is the same amount of FSH sufficient to allow normal ovulation in the breeding season when the ovary is stimulated by LH in a similar pattern to that which we have imposed. The results suggest rather, that the pulsatile LH pattern which was used in our anoestrous studies is not appropriate. McNatty et al (1984) have suggested that the failure of long term treatment of Romney ewes with LH pulses to maintain cyclical ovarian activity in anoestrus beyond two consecutive progestational phases may be due to an induced deficiency in FSH. However, no deficiency in FSH was apparent when LH injections were initiated since these resulted in ovulation with normal corpus luteum function. It can be assumed that the preovulatory surge of LH is released in response to an increase in endogenous oestradiol secretion from the developing follicle(s) (Baird et al., 1981). Therefore if it is assumed in Damline ewes that there is no change in sensitivity to the positive feedback effects of oestradiol in anoestrus, the earlier discharge of the preovulatory surge (29 $\pm$ 5 h after first LH injection) is some 25h earlier than after the start of the natural acceleration of pulsatile secretion of LH after prostaglandin-induced luteal regression in the normal oestrous cycle (Table 4). Since only LH has been injected there is no possibility of a direct effect of injected hormone on the sensitivity of the pituitary to release the preovulatory surge. Thus there is a clear difference between the response of Damline ewes and that of Finn-Merino ewes where the times to the onset of the preovulatory LH surge are the same in anoestrus and the breeding season.

One possible explanation for enhanced oestradiol secretion is that there are more follicles potentially available for stimulation in the anoestrous Damline than Finn-merino ewe. Thus LH alone would stimulate relatively greater amounts of oestradiol secretion leading to an earlier discharge of LH. In some way in support of this possibility is the response in Welsh ewes where fewer follicles would be available to respond to LH alone leading to a longer time period for oestradiol to increase

74

sufficiently to elicit a preovulatory surge. Indeed, with LH injections
alone, the delay to onset of the LH surge is similar to that in the
breeding season. When FSH is added, which might be expected to enhance
follicular development (Wright et al., 1981) and oestradiol secretion, the
preovulatory LH surge occurs some 29 h earlier, at a time comparable to
that seen in the LH treated Damline ewes.

However, these speculations do not resolve the question of why, when
ovulation occurs, as it appears to in all Welsh and Damline ewes with LH
injections with or without FSH only 3 of 10 Welsh ewes and 4/9 Damline
ewes have normal corpus luteum function. Normal corpus luteum function
was not correlated with time to preovulatory LH surge, or the amount of LH
released during the surge even though this was lower than the amount of LH
normally released in the oestrous cycle during the breeding season. What
is clear is that the continued injection of LH at hourly intervals from
48h onwards is not required to induce the preovulatory LH surge.

The present results suggest that the development of preovulatory
follicles in anoestrus is limited by the infrequent pulsatile secretion of
LH. The response of different breeds of ewe to pulsatile injections of LH
is probably related to the stage of anoestrus and the consequences of this
on the degree of follicular development and maturation at the time of
injection.

REFERENCES

Baird, D.T. 1978. Pulsatile secretion of LH and ovarian estradiol during
the follicular phase of the sheep oestrous cycle. Biol. Reprod. 18,
359-364.

Baird, D.T., Swanston, I.A. and McNeilly, A.S. 1981. Relationship between
LH, FSH and prolactin concentration and the secretion of androgens
and estrogens by the preovulatory follicle in the ewe. Biol.
Reprod. 24, 1013-1025.

McNatty, K.P., Hudson, N., Gibb, M., Ball, K., Fannin, J., Kieboom, L and
Thurley, D.C. 1984. Effects of long-term treatment with LH on
induction of cyclic ovarian activity in seasonally anoestrous ewes.
J. Endocr., 100, 67-73.

McNeilly, A.S. and Land, R.B. 1979. Effect of suppression of plasma
prolactin on ovulation, plasma gonadotrophins and corpus luteum
function in LH-RH treated anoestrous ewes. J. Reprod. Fert., 56,
601-609.

McNeilly, A.S., O'Connell, M. and Baird, D.T. 1982. Induction of ovulation and normal luteal function by pulsed injections of luteinizing hormone in anestrous ewes. Endocrinology, 110, 1292-1299.

Martensz, N.D., Scaramuzzi, R.J. and Van Look, P.F.A. 1979. Plasma concentrations of luteinizing hormone and follicle stimulating hormone during anoestrus in ewes immunized against oestradiol-17B, oestrone or testosterone. J. Endocr., 81, 261-269.

Oussaid, B. 1982. Etudes de l'activite ovarienne et de sa stimulation pendant l'anoestrus saisonnier chez la brebis Ile-de-France. Diploma de Docteur de 3^e cycle, l'Universite Pierre et Marie Curie, Paris VI.

Scaramuzzi, R.J. and Baird, D.T., 1977. Pulsatile release of luteinizing hormone and the secretion of ovarian steroids in sheep during anestrus. Endocrinology, 101, 1801-1806.

Wallace, J.M., McNeilly, A.S. and Baird, D.T. 1984. Induction of ovulation in two breeds of seasonally anoestrus ewes with multiple injections of LH alone or in combination with FSH. J. Reprod. Fert. In press.

Walton, J.S., McNeilly, J.R., McNeilly, A.S. and Cunningham, F.J. 1977. Changes in blood levels of prolactin, LH, FSH and progesterone during anoestrus in the ewe. J. Endocr., 75, 127-136.

Wright, R.W., Bondioli, K., Grammer, J., Kuzan, F. and Menino, A. Jr., 1981. FSH or FSH plus LH superovulation in ewes following estrus synchronization with medoxyprogesterone acetate pessaries. J. Anim. Sci. 52, 115-118.

Yuthasastrakosol, P., Palmer, W.M. and Howland, B.E. 1977. Release of LH in anoestrous and cyclic ewes. J. Reprod. Fert., 50, 319-321.

INDUCED BREEDING IN ANESTROUS MILKING EWES
OF DAIRY BREEDS: COMPARISON OF NORGESTOMET,
MEDROXYPROGESTERONE AND FLUOROGESTONE IN TWO
REGIMES OF PMSG

T. Alifakiotis

Animal Production Department, School of Agriculture,
University of Thessaloniki, Greece

ABSTRACT

The present investigation compares the suitability of norgestomet versus medroxyprogesterone and fluorogestone in inducing breeding in anestrous lactating ewes. Six hundred lactating ewes of two local dairy breeds were randomly divided into three equal groups during the spring. Ewes in group I were treated with 60 mg of medroxyprogesterone and in group II with 40 mg of fluorogestone using impregnated polyurethane vaginal sponges for 14 days. Ewes in group III were treated with 1,3 mg of norgestomet in impregnated mini-implants, placed subcutaneously in dorsal pinnae of the ear. The day the sponges and implants were removed, the ewes in each group were randomly allocated in two subgroups, a and b, and received 500 I.U. and 1,000 I.U. of PMSG, respectively. Vasectomized rams were used for estrus detection and fertile rams hand mating. The percentage of estrus exhibition and subsequent mating was similar among groups (98 and 84%, subgroups Ia and Ib vs 100 and 100%, subgroups IIa and IIb vs 100 and 96%, subgroups IIIa and IIIb). The meantime of estrus exhibition ranged between $40.1h \pm 6.2h$ to $42.3h \pm 4.1h$ in IIIa and IIIb, $46.4h \pm 6.1h$ to $49.2h \pm 3.9h$ in Ia and Ib, and $48.3h \pm 6.1h$ to $50h \pm 4.6h$ in IIa and IIb, respectively. The lambing percentage ranged between 59 to 49% in IIIa and IIIb, 62 to 39% in Ia and Ib, and 71 to 28% in IIa and IIb, respectively. The percentage of subsequent estrus returns was ranged between 39 to 50% in IIIa and IIIb, 38 to 59% in Ia and Ib, and 26 to 70% in IIa and IIb, respectively. The lambing percentage of the returns ranged in the same level (86 to 100% in group I vs 92 to 96% in group II vs 91 to 90% in group III, respectively) as well as the prolificacy (178 to 209 in group I vs 187 to 176 in group II vs 200 to 188 in group III, respectively). It was concluded that induction of breeding in anestrous lactating ewes of dairy breeds, using implants containing only 1,3 mg norgestomet, might be a useful tool to improve efficiency of sheep production.

INTRODUCTION

The current interest in avoiding any kind of hormonal compounds in livestock products causes problems for the producers regarding hormone utilization in the production processes. Thus, it is important that sheep milk used for human consumption contain no hormones at all or undetectable traces at least. In dairy breeds of sheep, whose milk is used for human consumption, must be effectively bred by using the minimum amount of these compounds.

The present research was, therefore, undertaken to establish a hormonal treatment for successful breeding of milking anestrous ewes by utilizing a minimal amount of the progestin norgestomet and by evaluating the efficacy of this treatment compared to commonly used progestines such as fluorogestone (cronolone) and medroxyprogesterone (MAP). Simultaneously, all the progestin treatments were tested under two regimes of PMSG.

MATERIALS AND METHODS

Six hundred lactating ewes of two native dairy breeds (Chios and Skopelos) were used in this study. All had lambed during the winter, and since then had been housed next to rams in natural daylength and temperature conditions until the middle of spring when the experiment was begun. They were subjected to standard housing and nutrition management and milked twice daily by hand. The ewes were randomly assigned to three experimental groups of 200 animals each. Ewes in group I were treated with 60 mg of medroxyprogesterone (MAP) in impregnated polyurethane vaginal sponges for 14 days. Ewes in group II were treated with 40 mg of fluorogestone (cronolone) in impregnated polyurethane vaginal sponges for 14 days. Ewes in group III were treated with 1.3 mg of norgestomet in impregnated cylindrical mini-implants (10 mm length, 3.3 mm diameter) from silicone, which were placed subcutaneously in the dorsal pinna of the ear, for 14 days. Immediately after withdrawal of sponges and implants, the ewes in each group were randomly allocated in two subgroups, a and b, and received 500 I.U. PMSG per ewe in subgroup a or 1,000 I.U. PMSG per ewe in subgroup b, respectively. The experimental desing is illustrated in table 1. Vasectomized rams were introduced to each group 24 hours following sponge or implant removal and were checked hourly for estrus exhibition. Ewes in estrus were led to the fertile ram for hand mating. All mated ewes were then placed in nearby pens with rams of proven fertility for free mating. Sponge or implant losses, estrus exhibition and matings, time of estrus exhibition, lambing %, prolificacy, and estrus returns were monitored and analysed statistically by using simple Chi-square comparisons (1).

RESULTS AND DISCUSSION

The loss of MAP-sponges from the vagina of the ewes in group I, during the period of 14 days, ranged from significant ($P<0.05$, subgroup Ia) to very significant ($P<0.01$, subgroup Ib) compared to Cronolone-sponges and to norgestomet implants, as shown in table 2, where the results of the present experiment are illustrated. The increased losses in ewes of group I were results of the shape, size, consistency, and surface texture of the sponges. This brings up an important point for consideration of sizing, shaping, and surface smoothness of sponges according to the vaginal anatomical structure of the sheep breed where they will be used.

The percentage of estrus exhibition and subsequent mating after hormonal treatment was similar between ewes treated with norgestomet (96 and 100%), medroxyprogesterone (94 and 98%), and fluorogestone (100 and 100%), respectively. Similar responses to similar hormonal treatments were reported for medroxyprogesterone and fluorogestone (2,3,4,5,6,7), however, for norgestomet similar responses had been succeeded by using higher doses (8, 9,10).

TABLE 1. Experimental desing for hormone-induced breeding in anestrous lactating ewes of dairy breeds, during spring, by using 3 different progestagens and two regimes of PMSG.

No. of group	Hormonal treatment					No. of ewes
	Type of hormone	Mean of administration	Duration of treatment	Days of treatment	Amount administ.	
Ia	MAP[1] PMSG[2]	sponge i.m.	14 days once	1st - 14th 14th	60 mg 500 IU	100
Ib	MAP PMSG	sponge i.m.	14 days once	1st - 14th 14th	60 mg 1000 IU	100
IIa	CRONOLONE[3] PMSG	sponge i.m.	14 days once	1st - 14th 14th	40 mg 500 IU	100
IIb	CRONOLONE PMSG	sponge i.m.	14 days once	1st - 14th 14th	40 mg 1000 IU	100
IIIa	NORGESTOMET[4] PMSG	implant i.m.	14 days once	1st - 14th 14th	1,3 mg 500 IU	100
IIIb	NORGESTOMET PMSG	implant i.m.	14 days once	1st - 14th 14th	1,3 mg 1000 IU	100

1. Medroxyprogesterone. Trade name "Repromap", Upjohn Co., Greece.
2. Pregnant Mare Serum. Trade name "Gestyl", N.V. Organon - Niadas, Greece.
3. 17a-acetoxy-11β-methyl-19-nor-preg-4-ene-3,20-dione. Trade name "Norgestomet", INTERVET Co., Angers, France.
4. Fluorogestone. Trade name "Chrono-gest", Searle Co., France.

TABLE 2. Estrus exhibition, mating, lambing and prolificacy in anestrous lactating ewes of dairy breeds after hormone-induced breeding during spring.

No of group (1)	No of ewes	After hormonal treatment					Returns and new estrus		
		Sponge or implant losses (p.100)	Estrus and mating (p.100)	Meantime of mating (X+SE, hours)	Lambing (p.100)	Prolificacy	Estrus and mating (p.100)	Lambing (p.100)	Prolificacy
Ia	100	13^a	98	50 ±4.8	62^a	221	38^a	86	178
Ib	100	16^k	94	48.3±6.1	39^{bE}	194	59^b	100^F	209
IIa	100	0^{bl}	100	46.4±6.1	71^c	217	26^a	92	187
IIb	100	1^{bl}	100	49.2±3.9	28^{dE}	191	70^b	96^F	176
IIIa	100	2^{bl}	100	40.1±6.2	59^A	189	39	91^B	200
IIIb	100	3^{bl}	96	42.3±4.1	49^E	201	50	90^F	188

1. See Table 1 for description of treatments.

* $P < 0.05$ for comparison a to b, c to d and e to f (same column) and A to B (different column).

** $P < 0.01$ for comparison k to l (same column) and E to F (different column).

The mean time of estrus exhibition and mating after removal of sponges and implants ranged between 40.1h ± 6.2h to 42.3h ± 4.1h in ewes treated with norgestomet and 46.4h ± 6.1h to 49.2h ± 3.9h and 48.3h ± 6.1h to 50h ± 4.6h in ewes treated with medroxyprogesterone and fluorogestone, respectively. Similar results for norgestomet treated ewes were reported by Cognie et al.

Immediately after the hormonal treatment, the lambing percentage from the matings on the induced estrus, in ewes which received norgestomet by mini-implants (group III), ranged between 59% when treated with 500 I.U. PMSG (subgroup IIIa) and 49% with 1,000 I.U. PMSG (subgroup IIIb). This suppression of the lambing percentage, as a result of increasing PMSG dose from 500 I.U. to 1,000 I.U., was not significant, statistically. However, the suppression of lambing percentage was significant in ewes which were treated with medroxyprogesterone (62% with 500 I.U. PMSG vs 39% with 1,000 I.U. PMSG) and fluorogestone (71% with 500 I.U. PMSG vs 28% with 1,000 I.U. PMSG). Embryo losses, as a result of high doses of PMSG were reported by Robinson (6) and recently by Evans and Robinson (11), who attested to extremely high levels of plasma estrogen proceeding ovulation and suggested that plasma estrogen must be within a restricted range for optimum fertility. In fact, it is known that estrogen has quantitative effects on sperm transport and survival in the female genital tract (12) and the relative survival of ova is markedly reduced as ovulation rates increase (11). Further, regarding breeding of induced anestrous ewes after progestins and PMSG Evans et al. found that the breeding efficiency increases with increasing doses of PMSG only up to 800 I.U., above which the effects of PMSG are counter-productive (11). Recently, the use of norgestomet implants in Hampshire ewes for regulation of estrous cycles (13) showed that when the dose was increased from 3 mg to 6 mg, the number of spermatozoa in uteri and oviducts reduced significantly.

In previous reports concerning the use of norgestomet in sheep, Cognie et al. (9) found that 3 mg of this progestin used as mini-implants for 12 days followed by 500 I.U. of PMSG resulted in a lambing percentage of 66.7%. The treatment was given in late spring and early summer. The beginning of breeding season in Aragonesa sheep (9) in a similar study, the use of 4.5 mg norgestomet (3 mg by mini-implants and 1.5 mg by injection) with 750 I.U. of PMSG at implant removal, resulted in a lambing percentage of 43% during May and 57% during June in non-lactating ewes of Rambouillet breed (8). These two reports regarding the use of norgestomet in anestrous ewes are comparable to breeding effectiveness reported in the present experiment, although the dose of norgestomet in the later reports was twofold (9) to three-fold (8) compared to the present study. The prolificacy from the matings on the induced estrus reached similar levels in ewes of all groups and no significant differences were detected among groups and subgroups.

Further, the percentage of subsequent returns to estrus and new exhibitions of behavior showed significant differences between subgroups treated with medroxyprogesterone (38% with 500 I.U. PMSG vs 59% with 1,000 I.U. PMSG) and fluorogestone (26% with 500 I.U. PMSG vs 70% with 1,000 I.U. PMSG), while no difference was detected between subgroups in ewes treated with norgestomet. The differences were due to conception variability in estrus induced subgroups. Furthermore, the lambing percentage of the estrus returns was restored (86% to 100% in group I vs 92% to 96% in group II vs 91% to 90% in group III, respectively) and the progestational suppression of fertility during the induced estrus was further deleted. The prolificacy from the matings of the returns and new estrus ranged in the same level among subgroups and was similar to the induced estrus matings.

CONCLUSION

Successful induction of breeding and reproductive performance in anestrous milking ewes of dairy breeds can be obtained using either medroxyprogesterone of fluorogestone or norgestomet; however, the used amount of
the latter progestin was limited only to 1.3 mg. The suppression of fertility from possible use of high doses of PMSG, such as 1,000 I.U., at the
end of the progestional treatment is negated by using norgestomet. Thus,
induction of breeding in anestrous milking ewes of dairy breeds using implants containing only 1.3 mg of norgestomet might be a useful tool to
improve efficiency of sheep production.

REFERENCES

1. Ostle B. Statistics in research. Iowa State Univ. Press (1969).

2. Alifakiotis, T., Matsoukas, J., Hatjiminaoglou, J., and N. Zervas. Induced breeding in anestrous milking ewes by using twice-repeated
   PMS combined with estradiol - 17β. Ann. Biol. anim. Bioch. Biophys.
   18 (2A): 229-235 (1978).

3. Cognie, Y., Cornu, C., and P. Mauleon. The influence of lactation on
   fertility of ewes treated during post-partum anoestrus with vaginal
   sponges impregnated with FGA. Proc. Symp. Physiopath. Reprod. art.
   Ins. p. 33-36 (1974).

4. Margaritis, I., Samouelidis, S., and V. Semadopoulos. Conception rate
   in sheep following oestrus synchronization by means of "silestrus"
   implants, map sponges and injection of PMS, during anoestrus. Proc.
   Symp. Physiopath. Reprod. art. Ins. p. 58-65 (1974).

5. Rahman, S.S., and W.D. Kitts. Hormonal control in lactating and non-
   -lactating anoestrus ewes. Can. J. Anim. Sci. 47:65-69 (1967).

6. Robinson, T.J. The control of fertility in sheep. Part II. The augmentation of fertility by gonadotrophin treatment of the ewe in the
   normal breeding season. J. Agric. Sci. Camb. 41:6-63 (1951).

7. Thimonier, J., Mauleon, P., Cognie, Y., ang R. Ortavant. Déclenchement
   de l' oestrus et obtenion de la gestation pendant l' anoestrus post-
   -partum chez la brebis à l' aide d' éponges vaginales impregnées d'
   ácetate de fluorogestone. Ann. Zootech. 17:257-273 (1968).

8. Carpenter, H.R. and J.C. Spitzer. Response of anestrous ewes to norgestomet and PMSG. Theriogenology 15:389-393 (1981).

9. Cognie, Y., Folch, J., and M. Alonso de Miguel. Utilization des implants sous-cutanés de SC 21009 pour la synchronization des chaleurs
   chez la brebis (personal communication) (1980).

10. Spitzer, C.J., and R.H. Carpenter. Synchronized breeding of cycling
    ewes to produce fetuses of known gestational age. Lab. Anim. Sci.
    29: 755-758 (1979).

11. Evans, G., and T.J. Robinson. The control of fertility in sheep: endo-
    crine and ovarian responses to progestagen-PMSG treatment in the
    breeding season and in anoestrus. J. Agric. Sci. Camb., 94: 69 - 88
    (1980).

12. Robinson, T.J. Factors involved in the failure of sperm transport and survival in the female reproductive tract. J. Reprod. Fertil. (suppl.) 18: 103-109 (1973).

13. Feccia, C.R., Riesen, W.J., and C.O. Woody. Sperm numbers in progestin treated ewes. J. Anim. Sci. 51 (suppl.): 143 (Abstr.) (1980).

DISCUSSION

Chairman: Lamming, E. (U.K.)

Seasonal anestrus in sheep is characterized by an LH episode frequency which is below that necessary for the induction of the final phases of follicle growth. In some breeds the restoration of this inadequacy by appropriate therapy with either purified LH or GnRH will promote ovulation. However, in "deeply" anestrous breeds it is necessary to use a combination of FSH and LH. Such differences indicate that the endocrinology of seasonal anestrus is not a consistent phenomenon and depth of anestrus may relate to differences in the extent of early follicle development. However, the ovaries are not quiescent during the non-breeding season, even in the very seasonal breeds. Indeed, follicle recruitment up to a fairly advanced stage of development is present. Genetic influences are manifest, in that the mechanisms controlling ovulation rate still appear to be present even during seasonal anestrus.

Although it has been known for many years that altering the light:dark ratio can be used to manipulate the seasonal nature of reproduction, more recent data indicate that there is a critical period in relation to the start of the light phase when light must be present for sheep to read a "long day", but this can be achieved by providing a 1 hour pulse of light in association with a "short day". A consistent conclusion from several studies was that the seasonal change in prolactin was not of an important component in the regulation of seasonal breeding. Although earlier studies suggested that the secretions of the pineal gland were not necessary for seasonal breeding, recent results show that the seasonal pattern of breeding can be manipulated by administration of melatonin, although the precise neuroendocrine mechanisms involved are still largely unknown.

SESSION II

SEASONAL ANESTRUS IN THE SOW AND MARE

Chairman: J.F. Roche

STIMULATION OF OVARIAN ACTIVITY IN THE PONY
MARE DURING WINTER ANOESTRUS

B. Bour, E. Palmer, M.A. Driancourt
Institut National de la Recherche Agronomique
Station de Physiologie de la Reproduction
Nouzilly, 37380 Monnaie, France

ABSTRACT

Two types of stimulation of winter ovulatory inactivity are described :
1°) long term stimulation (70 days) with photoperiodic treatments, the photo-inducible phase of which is defined as one hour's light, nine hours and a half after dusk;
2°) short term stimulation (10 days)with LH and FSH gonadotropins, the sensitivity of the mare to these resulting in the division of the transition period in early (FSH+LH stimulatory) and late (LH stimulatory) transition. The link between the two means of hastening the ovulatory season is to be found in the transmission mechanism from light to the pituitary through the pineal.

INTRODUCTION

As many other species, the mare is a seasonal breeder; the fertile period occurs in the springtime, mares foaling after a pregnancy of nearly a year (11 months), during the same appropriate phase of the annual cycle (climate and food availability) (KARSCH et al., 1984). The frequency and the length of ovulatory inactivity, known as anoestrus, in winter, depend on different factors:

- the breed : more pony mares than saddle mares show an absence of ovarian activity in winter (100 %, n=57 vs 66 %, n=8; PALMER, 1978) and cyclicity resumes later in pony mares than in saddle mares. (May 6th, n=12 vs April 1st, n=12) in similar conditions (GINTHER, 1979).

- the individual tendency: winter inactivity is more frequent in mares which showed inactivity the preceding year than in mares which did not (78 %, n=18 vs 33 %, n=12; PALMER, 1978).

- the age: all 18-month-old fillies have inactive ovaries during the winter preceding their first reproductive season (n=12; BOUR, DRIANCOURT, PALMER, 1984),

- the physiological state during the preceding summer: mares having foaled and lactated will most probably be inactive the next winter (100 % vs 30 % for non previously lactating mares ; PALMER and DRIANCOURT, 1982),

- body condition: a long-term effect of nutrition seems to monitor ovarian activity; mares in poor body condition will be in winter anoestrus even after several months of proper feeding; initiation of cyclicity can be as late as June or July (see below Exp. V).

Seasonal inactivity is defined in the mare as in other species as the absence of terminal follicular growth and ovulation. Yet activity of the pituitary is different from what can be observed in the sheep : LH is low throughout the winter, but no difference in FSH levels can be shown between anoestrus and breeding season in the intact mare (GINTHER, 1974). This originality is particularly clear when considering LH variations in ovariectomized females over a year. The mare has low LH levels in winter, high levels in summer (FREEDMAN et al., 1979), whereas no variations exist in the ovariectomized ewe (maximum LH secretion throughout the year), which needs oestradiol feedback to lower circulating LH to undetectable level during the summer anoestrus period (LEGAN, KARSCH, FOSTER, 1977). FSH follows the same pattern as LH in the ovariectomized mare, indicating depressed pituitary activity during winter.

FREEDMAN et al. (1979) observations demonstrate that the secretion of gonadotropins by the pituitary is a fonction of day-length :
the spring increase in gonadotropins of ovariectomized mares takes place two months earlier when mares are lighted in "long days" (16 hours light) from December on (Text-Figure 1).

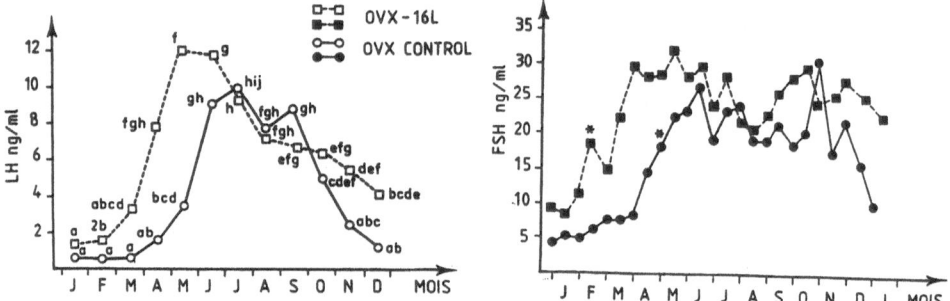

Fig. 1 Comparison of changes in mean monthly LH and FSH profiles in Control-photoperiod (———) and 16 h daily photoperiod (−−−−) ovariectomized mares (from FREEDMAN et al., 1979).

BUCKARDT (1947) had observed the stimulatory effect of long days, first revealing the possible long-term control of the first ovulation in the year with 2 months photoperiodic treatment.

A short-term response is attempted with gonadotrophic treatments.

PHOTOPERIODIC STIMULATION. A LONG-TERM RESPONSE

Photoperiodic treatments (16 h light : 8 h dark from November 25th) have been used for producing an early breeding season, beginning after around 70 days of treatment, that is in the first days of February (PALMER, 1979). The experiments described below intended to define a "photosensitive phase", i.e. a period of time within a 24 h day where the presence of light would stimulate ovarian activity in the mare, and to confirm the delay to first ovulation.

Conditions of feeding, stabling, and artificial lighting are precisely described in PALMER, DRIANCOURT, ORTAVANT (1982). Lighting programmes, started at the end of November, are formulated as hours of light (nL) from dawn to dark, the latter symbolized as a dark strip.
Jugular blood was collected twice a week and assayed for progesterone concentration in order to assess date of first ovulation and further cyclicity (PALMER et JOUSSET, 1975). The same end-point, the date of the first ovulation in the year detected by progesterone rise, measured in weeks since January 1st, was used in all experiments.

Experiment I was designed further to investigate the minimum day-length which can stimulate ovarian activity. 35 Welsh pony mares were allocated to 7 experimental groups shown in table 1 with the corresponding mean date of first ovulation.

TABLE 1   Effect of day-length on resumption of ovarian activity.

EXPERIMENT I

LIGHT PROGRAMME	INTERVAL FROM JANUARY 1st TO FIRST OVULATION (WEEKS)
CONTROL PHOTOPERIOD	$18,0 \pm 0,8$
8 L	$19,7 \pm 3,3$
11,5 L	$15,4 \pm 2,5$
13 L	$13,0 \pm 1,9$
14,5 L	$4,9 \pm 1,1$ [*]
16 L	$5,7 \pm 1,0$ [*]
20 L	$11,5 \pm 2,5$

n = 5 per group.

When comparing different groups to Control, only 14.5L and 16L were stimulatory. If artificial photoperiodic stimulation of 14.5L is applied, a time-lag of 2 months to ovulation is necessary. By contrast, mares in control groups presented their first ovulation 18 weeks after January 1st, around May 1st. At this point, natural photoperiod just reaches 14.5L. In addition, mares kept under 8L begin cycles approximately at the same time. Thus, under natural conditions, the increase in ovarian activity is probably not a consequence of photoperiodic stimulation, which may be only a synchronizer to an endogenous circannual rhythm.

Experiment II, like Experiment III, tested the effect of asymmetrical skeleton photoperiod, looking for a "photoinducible phase", assuming that light given at the right time of the darkness period would be sufficient to induce stimulation of ovarian activity.

In minimum stimulatory photoperiod (14.5L), superposition of photosensitive phase and light period should be limited at one extremity of the light period, i.e. at dawn or dusk. These two possibilities were tested in asymmetrical skeleton : 8L and 1L ending 14.5 hours later if photoinduction occurs at dusk; 8L and 1L ending 18.5 hours later (i.e. 14.5 hours before the end of the main 8 hours of light) if photoinduction occurs at dawn.

The group receiving 8 L and 1 L ending 16.5 hours after dawn could be read both ways as 16.5 L followed by 8 hours dark.

On december 1st 16 Welsh pony fillies were randomly divided into four groups summarized in the table below with corresponding results.

TABLE 2   Effect of asymmetric skeleton photoperiods on stimulation of early ovulation.

EXPERIMENT II

LIGHT PROGRAMME	INTERVAL FROM JANUARY 1st TO FIRST OVULATION (WEEKS)
CONTROL PHOTOPERIOD	14,1 ± 1,7
8 L 1L ←— 14,5 hours —→	14,5 ± 1,4
8 L 1L ←—16,5 hours—→	10,3 ± 2,4
8 L 1L ←—— 18,5 hours ——→	7,0 ± 2,9 *

n = 4 per group.

These results suggest a stimulation in the group receiving 8L and 1L ending 18.5 hours after dawn, assuming that photoinduction occurs at dawn, 9.5 hours after dusk.

Experiment III was designed further to investigate this hypothesis. 25 anoestrous adult Welsh pony mares were randomly allocated to 5 experimental groups, 4 of which served only as negative or positive controls, or as confirmation of groups of Experiment II; the group receiving 4L and 1L ending 14,5 hours after dawn, testing the position of the photosensible phase 9.5 hours after dusk, gives a positive answer as shown in table 3.

TABLE 3   Effect of 1 h light at different times after dusk on date of first ovulation.

EXPERIMENT III

LIGHT PROGRAMME	INTERVAL FROM JANUARY 1st TO FIRST OVULATION (WEEKS)
CONTROL PHOTOPERIOD	$15,1 \pm 0,9$
16 L	$6,9 \pm 1,9$ [*]
8 L ────1L── ──18,5 hours──	$7,2 \pm 0,8$ [*]
8 L ──1L── ──14,5 hours──	$22,7 \pm 1,4$
4 L ──1L── ──14,5 hours──	$6,3 \pm 1,8$ [*]

n = 5 per group.

This "4L+1L" group also clearly demonstrates that the total number of hours of light is not essential for perception of "long days".

From these 3 experiments a long day can be defined in the mare as the presence of light about 9.5-10.5 hours after the beginning of night. In contrast, the total number of hours of light is of little importance, provided a minimum length of darkness is given. The photoperiodic "long-day" artificial stimulation results in an early cyclicity about 2 months after the beginning of treatment.

GONADOTROPHIC STIMULATION A SHORT TERM RESULT

Supplementation in FSH and LH (pituitary extracts) results in multi-ovulation of anoestrous mares (DOUGLAS, NUTI, GINTHER, 1974). However, a reliable treatment to induce single ovulations is not available. Mc NATTY et al., (1981), Mc NEILLY, O'CONNEL and BAIRD (1982) were successful in inducing ovulation by injecting or infusing LH to anoestrous ewes. Moreover, several observations - no rise in LH after the last ovulation of the season (SNYDER et al., 1979), progressive increase in the LH peaks' maximum levels, associated with the first ovulations of the season (OXENDER et al., 1977), much depressed LH but not FSH when mares do not show a post-partum ovulation (PALMER, DRIANCOURT, 1983) - led to the hypothesis that LH was the limiting factor for seasonal initiation of cyclicity in the mare.

Experiment IV tested the effect of an LH supplementation using daily low doses of hCG during February and March on follicular maturation, ovulation and luteal function. hCG treatments of 24 selected anoestrous pony mares were designed to mimick an increase in LH levels before, during, and after ovulation in a factorial design experiment :
1) PRE-treatment during February (n=8) or March (n=8) with daily i.m. injections of 200 I.U. (max 21 days) or Control (n=8).
2) PEAK-treatment (Control n=12, treated n=12) with 2 000 I.U. i.v. on the day a preovulatory follicle was observed (30 to 35 mm).
3) POST-treatment (Control n=12, treated n=12) with 200 I.U. i.m. daily for 5 days following ovulation.
Follicular growth and early pregnancy were assessed by echographic examination of the ovaries and uterus. Progesterone, total estrogens, FSH and LH were assayed as previously described on daily samples. Levels of hCG antibodies were estimated using the method of THOMAS et al. (1972).
Statistical analysis was realised by non parametrical methods, mainly U test. Comparisons were made between groups of mares according to positive or negative responses to the treatments. Means are given ± s.e.m. In February, PRE-treatment induced a follicle of 30 mm in 2 mares and ovulation in only one mare but was highly successful in March (7/8 mares ovulated after 8.8±2.4 days vs 1/7 controls; P<0.05). The size of the largest follicle observed on the first day of treatment was bigger in March, 20.4±2.0 mm vs 14.4±1.5 mm in February, indicating an increase in

follicular activity. Follicles ovulating after PRE-treatment were smaller than spontaneously ovulating follicles (35.6+0.7 mm vs 39.3+1.5 mm; P<0.01) and produced less estrogens (0.6+0.1 ng/ml vs 1.3+0.3 ng/ml; P<0.01).

PEAK-treatment was not necessary for ovulation of a PRE-induced follicle (3/3 with PEAK vs 6/6 without) but urged on the ovulation of spontaneous follicles (with PEAK 4/4, without 1/4 ovulating in February or March).

Mean day 1 to day 5 post-ovulation Progesterone levels were not different between POST-treated animals and controls.

Injections of hCG didn't alter gonadotropin profiles during treatment but:

1) Mares ovulating after PRE-treatment did not show a preovulatory LH rise,

2) Mares unsuccessfully treated in February developed antibodies to hCG, associated with a refractoriness to the PEAK injection and showed subsequently higher LH levels before and after first ovulation than untreated mares,

3) The FSH pattern didn't differ between mares.

Fertility of PRE-induced ovulations was as good as that of ovulations of spontaneous follicles. The hypothesis of a lack of LH in March, but not FSH, for producing a first fertile ovulation 25 days earlier than Control, is confirmed by the experiment and leads to the following diagram of the transition period.

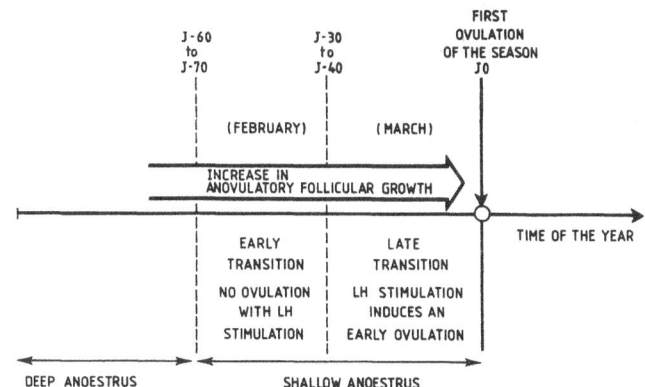

Fig. 2   Characterization of transition period to ovulatory season.

The high amounts of LH assayed in the RIA might either be effective circulating concentrations, or more probably LH linked to the antibodies in a complex, this complex being dissolved during the assay. Though antibodies don't hinder further cyclicity, the refractoriness to hCG, already described by LOY (1966), but denied by ROSER et al., (1979) - that they induce, makes the treatment unusable in current practice.

Experiment V intented to investigate the transition period further, by
- testing the association of FSH to LH in February,
- trying to obtain single ovulation by changing the FSH/LH ratio halfway in the stimulation,
- estimating the dosage of equine LH equivalent to 200 I.U. of hCG, used in Exp. IV,
- characterizing ovarian activity after the induced ovulation.

Complete pituitary extracts (E_{FSH+LH}) were prepared using the technique of GUILLOU and COMBARNOUS (1982), FSH activity being dissociated at pH 5,5 afterwards to obtain pure LH activity (E_{LH}). Mares were injected i.m. until ovulation or during 15 days at most exactly at the same periods of the year as in Exp. IV. Experimental design is summarized in table 4.

TABLE 4   Experimental design (Exp. V).

FEBRUARY			MARCH		
treatment	dose per day	number of mares	treatment	dose per day	number of mares
FEBRUARY CONTROL		n = 5			
$E_{FSH + LH}$	20 mg	n = 5			
$E_{FSH + LH}$ then E_{LH}	20 mg 20 mg	n = 5			
E_{LH}	20 mg	n = 5	E_{LH}	20 mg	n = 5
			E_{LH}	12 mg	n = 5
			E_{LH}	6 mg	n = 5
			hCG	200 UI	n = 5
			MARCH CONTROL		n = 5

The "E_{FSH+LH} then E_{LH}" animals receive complete extracts until 3 criteria are obtained together:
- they have received E_{FSH+LH} during a minimum of 4 days,
- the size of the largest follicle (F) is over 22 mm,
- its growth since the beginning of treatment exceeds 7 mm.

Observations, assays and analysis techniques are the same as described in Exp. IV.

Treatments induced a significant increase in FSH and LH for complete extracts $(E_{FSH + LH})$ in LH only for treated extracts (E_{LH}), in spite of a residual 20 % FSH activity in these.

February stimulation results are presented in table 5.

TABLE 5   February stimulation results.

TREATMENT GROUP	Number of mares treated	Number of mares ovulating	F size beginning of treatment (mm)	Interval to ovulation (days) Mean	Range	Number of ovulations
FSH + LH	5	4	12.4 + 4.4	9.3 + 4.3	2 - 13	2
FSH + LH then LH	5	5	11.8 + 3.6	8.6 + 1.5	6 - 10	1
LH	5	1	10.8 + 4.0	10	/	0
CONTROL	5	0	14.0 + 2.8	/	/	/

Duration of treatment to ovulation is the same as after hCG stimulation the preceding year. Two double ovulations occurred in the "FSH+LH" group, and only one in the "FSH+LH then LH" : the 20 mg dose seems to be adequate and the succession of LH to FSH+LH successful in obtaining a single ovulation.

This ovulation was not followed by other cycles. Mares resumed cyclicity at the same time as Control.

The mare that did not ovulate was 16 months old and had not reached puberty at the time of the experiment because of improper feeding in its young age.

Before treatment, gonadotropin levels didn't differ between February and March, but size of the largest follicle (F) was significantly higher in March (16.8+1.1 mm vs 12.2+0.9 mm; P<0,01) as plasmatic estrogen levels (0.5+0.1 vs 0 ng/ml; P<0.01).

TABLE 6   March stimulation results.

TREATMENT GROUP	Number of mares treated	Number of mares ovulating	F size beginning of treatment (mm)	Interval to ovulation (days) Mean	Range	Number of ovulations
LH 20	5	1	20.6 + 5.6	12	/	0
LH 12	5	4	20.4 + 5.4	11.8 + 5.9	4 - 20	0
LH 6	5	2	15.4 + 4.2	15.5 + 3.5	12 - 19	0
hCG	5	1	13.4 + 5.9	13	/	0
CONTROL	5	0	14.2 + 3.9	/	/	/

94

As shown in table 6, only 8 mares out of 20 ovulated after 12.9±5.5 days of LH supplementation, 5 of which went on cycling. Two mares in the hCG treatment group had higher levels of antibodies and were refractory. The other non ovulating mares in the four stimulated groups had a larger follicle of a reduced size at the beginning of treatment (15.0±1.2 vs 20.6±2.0 mm; P<0.01), and ovulated much later in the year than Control mares of Exp. IV (132.4±7.4 vs 94.0±8.1 days).

They were mares in poor body-condition after improper feeding during the preceding months; their transition period came in late May or June and they were unable to respond to an LH stimulation in March, as shown in figure 3.

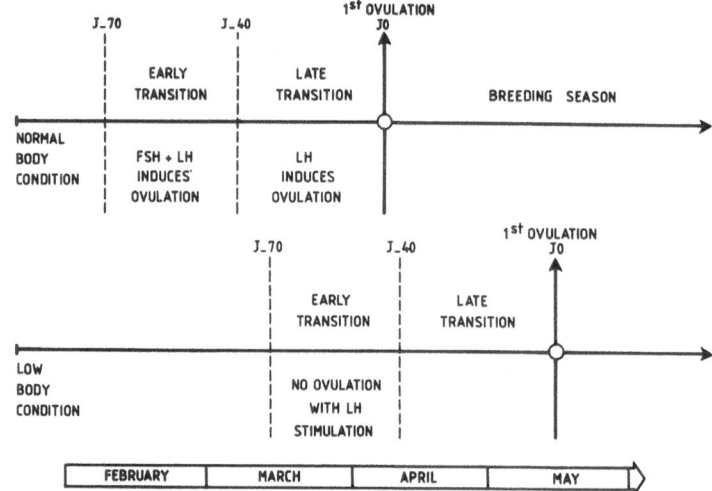

Fig. 3   Variability in the transition phases and the sensitivity to FSH and LH stimulation.

Other pharmacological stimulations of ovarian activity have been experimented during the different phases of anoestrus. ALLEN et al., (1980) consider that progestagen treatment in late transition hastens the first ovulation of the year, but many trials failed to confirm this statement (PALMER, BOUR, CHEVALIER 1982; SCHEFFRAHN et al., 1982; THOMPSON et al., 1984); they conclude only to the synchronization effect in very early breeding season. PALMER (1984) tested the effect of antioestrogens (Tamoxifen) on inactive pony mares but didn't obtain any stimulation with the doses used.

EVANS and IRVINE (1979) produced ovulation with GnRH after progesterone impregnation of the mares, but establishment of luteal function remained questionable, in contrast with the results of Mc LEOD et

HARESIGN (1984) in the ewe. GnRH alone (EVANS and IRVINE 1979; ALLEN and ALEXEEV, 1980) on in association with oestradiol cyprionate (HENNINGTON et al., 1982) failed to stimulate inactive mares.

HART et al. (1984) examined seasonal variation in hypothalamic content of GnRH, and pituitary content of LH and FSH in the mare. GnRH and LH were lowest during midanoestrus, while no effect of season was observed on either GnRH receptors or FSH. These authors conjecture, agreeing with our results in Exp. IV, that the low concentration of LH available for release delays ovulation during the period of transition into the breeding season. As for FSH concentrations in the pituitary, they are consistent with measured plasmatic FSH levels. The need for exogenous FSH for deep anoestrus and early transition stimulations (Exp. IV et V) leads to the hypothetis that the FSH detected in winter might be biologically inactive. New Zealand workers (ALEXANDER and IRVINE, pers. comm.) are trying to investigate this field.

The link between photoperiod and gonadotrophic stimulation has been approached through studies on the pineal and melatonin GRUBAUGH et al. (1982) demonstrated that pinealectomy blocks the ability of pony mares to respond to photostimulation and suppresses time-trends in plasma melatonin concentrations observed in intact mares (significant increase during darkness). WESSON et al., (1979) had established increased biosynthetic activity of the pineal during the late fall and winter at the time of reduced ovarian activity; this was consistent with the hypothesized antigonadal activity of the pineal in the mare. KILMER et al., (1982) demonstrated the circadian nature of melatonin secretion. The constant melatonin supplementation realized by THOMPSON, GODKE and NETT (1983) did not have any effect on ovarian activity; this could have been expected in a "long-day" breeder, in which ovarian activity is depressed by melatonin. In contrast, NETT and NISWENDER (1982) attempts in the ewe, a "short-day" breeder, were positive in hastening the onset of the ovulatory season in the fall or extending its length in the spring.

In addition, KARSCH et al., (1984) observed altered LH secretion after melatonin infusions in the ewe demonstrating the stimulatory effect of melatonin on the hypothalamus-pituitary axis with a time-lag of 30 to 50 days. This latency, resulting most probably from the link between the melatonin receptor and the LH pulse generator would explain the long-term response to photoperiodic stimulation, but these steps of the mechanism are still to be investigated in the mare.

REFERENCES

Allen, W.E. and Alexeev, M. 1980. Failure of an analogue of gonadotrophin releasing hormone (HOE 766) to stimulate follicular growth in anestrous pony mares. Equine Vet. J., 12 (1), 27-28.

Allen, W.R., Urwin, V., Simpson, D.J., Green-Wood, R.E.S., Crowhurst, R.C., Ellis, D.R., Ricketts, S.W., Hunt, M.D.N., Wingfielddigby, N.J., 1980. Preliminary studies on the use of an oral progestagen to induce estrus and ovulation in seasonally anestrous thoroughbred mares. Equine Vet. J., 12, (2), 141-145.

Bour, B., Driancourt, M.A. and Palmer, E. 1984. Problèmes liés à la mise à la reproduction avant le 1er mai. In press.

Buckardt, J. 1947. Transition from anoestrus in the mare and the effect of artificial lightning. J. Agri. Sci. Camb., 37, 64-68.

Douglas, R.H., Nuti and Ginther, O.J. 1974. Induction of ovulation and multiple ovulation in seasonally anovulatory mares with equine pituitary fractions. Theriogenology (1974), 2, (6), 133-142.

Evans, M.J., Irvine, C.H.G. 1977. Induction of follicular development, maturation and ovulation by GnRH administration to acyclic mares. Biol. Reprod., 20, 567-574.

Evans, M.J. and Irvine, C.H.G. 1979. Induction of follicular development and ovulation in seasonally acyclic mares using Gn-RH and Progesterone. J. Reprod. Fert., Suppl. 27, 113-121.

Freedman, L.J., Garcia, M.C. and Ginther, O.J. 1979. Influence of photoperiod and ovaries on seasonal reproductive activity in the mare. Biol. Reprod. 20, 567-574.

Ginther, O.J. 1974. Occurrence of anestrus, estrus, diestrus and ovulation over a 12 month period in mares. Am. J. Vet. Res., 35 (9), 1173-1179.

Ginther, O.J. 1979. Reproductive biology of the mare. Basic and applied aspects. Mc Naughton Gunn Inc. Ed., Anntlerbor, Mich.

Grubaugh, W., Sharp, D.C., Berglund, L.A., Mc Dowell, K.J., Kilmer, D.M., Peck, L.S. and Seamans, K.W. 1982. Effects of pinealectomy in pony mares. J. Reprod. Fert., Suppl 32, 293-295.

Guillou, F. and Combarnous, Y. 1983. Purification of equine gonadotropins and comparative study of their acid-dissociation and receptor-binding specificity. Biochim. Biophys. Acte 755, 229.

Hart, P.J., Squires, E.L., Imelk, J. and Nett, T.M. 1984. Seasonal Variation in Hypothalamic Content of GnRH, Pituitary Receptors for GnRH, and Pituitary Content of LH and FSH in the mare. Biol. of Reprod. 30, 1055-1062.

Hennington, D.L., Kreider, J.L., Potter, G.L., Harms, P.G. and Fleeger, J.L. 1982. The effect of GnRH on induction of follicular development and ovulation in anovulatory and ovulatory pony mares. J. Anim. Sci., 55, Suppl 1, 29 (Abstr. 71).

Karsch, F.J., Bittman, E.L., Foster, D.L., Goodman, R.L., Legan, S.J. and Robinson, J.E. 1984. Neuroendocrine basis of seasonal reproduction. Recent prog. horm. res., 40, 185-231.

Kilmer, D.M., Sharp, D.C., Berglund, L.A., Grubaugh, W., Mc Dowell, K.J. and Peck, L.S. 1982. Melatonin rythms in pony mares and foals. J. Reprod. Fert., Suppl. 32, 303-307.

Legan, S.J., Karsch, F.J. and Foster, D.L. 1977. The endocrine control of seasonal reproductive function in the ewe : a marked change in response to the negative feedback action of estradiol on Luteinizing Hormone secretion. Endocrinology, 101, 818-824.

Loy, R.G. and Hughes, J.P. 1966. The effects of human chorionic gonadotrophin on ovulation, length of oestrus and fertility in the mare. Cornell Vet., 56, 41-50.

Mc Leod, B.J. and Haresign, W. 1984. Induction of fertile oestrus in seasonally anoestrous ewes with low doses of GnRH. Anim. Reprod. Sci., 7, 413-420.

Mc Natty, K.P., Gibb, M., Dobson, C. and Thurley, D.C. 1981. Evidence that changes in luteinizing hormone secretion regulate the growth of the preovulatory follicle in the ewe. J. Endocr., 90, 375-389. .

Mc Neilly, A.S., O'Connel, M. and Baird, D.T. 1982. Induction of ovulation and normal luteal function by pulses of LH in anestrous ewes. Endocrinology, 110, 1292-1299.

Nett, T.M. and Niswender, G.D. 1982. Influence of exogenous melatonin on seasonality of reproduction in sheep. Theriogenology, 17 (6), 645.

Oxender, W.D., Noden, P.A. and Hafs, H.D. 1977. Estrus, ovulation and serum Progesterone, Estradiol and LH concentration in mares after an increased photoperiod during winter. Am. J. Vet. Res., 38, 203-207.

Palmer, E. 1978. Control of the estrous cycle of the mare. J. Reprod. Fert., 54, 495-505.

Palmer, E. 1979. Reproductive management of mares without detection of oestrous. J. Reprod. Fert., Suppl. 27, 263-270.

Palmer, E. 1984. Attempts to improve synchronization of ovulation and to induce superovulation for embryo collection in the mare. Havemeyer Equine Embryo Transfer Symp., Cornell, oct. 1984.

Palmer, E., Bour, B. and Chevalier, F. 1982. Pharmacological control of reproductive mechanisms in the equine female in "Veterinary pharmacology and toxicology", Ruckebusch et al. Ed., MTP Press, 1983.

Palmer, E. and Driancourt, M.A. 1983. Interactions of season of foaling, photoperiod and ovarian activity in the equine. Livestock Prod. Sci., 10, 197-210.

Palmer, E., Driancourt, M.A. and Ortavant, R. 1982. Photoperiodic stimulation of the mare during winter anoestrus. J. Reprod. Fert., Suppl. 32, 275-282.

Palmer, E. and Jousset, B. 1975. Urinary estrogens and plasma progesterone levels in non-pregnant mares. J. Reprod. Fert., Suppl. 23, 213-221.

Roser, J.F., Kiefer, B.L., Evans, J.W., Neely, D.P. and Pacheco, C.A. 1979. The development of antibodies to human chorionic gonadotrophin following its repeated injection in the cyclic mare. J. Reprod. Fert., Suppl. 27, 173-179.

Scheffrahn, N.S., Wiseman, B.S., Vincent, D.L., Harrison, P.C. and Kesler, P.J. 1982. Reproductive hormone secretions in pony mares subsequent to ovulation control during late winter. Theriogenology (1982), 17 (6), 571-585.

Snyder, D.D., Turner, D.D., Miller, K.F., Garcia, M.L. and Ginther, O.J. 1979. Follicular and gonadotrophic changes during transition from ovulatory to anovulatory season. J. Reprod. Fert., Suppl. 27, 95-101.

Thompson, P.L., Godke, R.A. and Nett, T.M. 1983. Efects of melatonin and thyrotropin releasing hormone on mares during the non breeding season. J. Anim. Sci., 56, 668-677.

Thompson, D.L., Revilles, S.T., Derrick D.J. and Walker, M.P. 1984. Effects of placement of intravaginal sponges on LH, FSH, estrus and ovarian activity in mares during the nonbreeding season. J. Anim. Sci., 58 (1), 159-164.

Wesson, J.A., Orr, E.L., Quay, W.B. and Ginther, O.J. 1979. Seasonal relationship between pineal Hydroxyindole-O-Methyltransferase (HIOMT) Activity and Reproductive Status in the Pony. Gen. Tomp. Endocr., 38, 46-52.

PRACTICAL CONTROL OF ANOESTRUS IN THOROUGHBRED BROODMARES

W.R. Allen

T.B.A. Equine Fertility Unit,
Animal Research Station,
307 Huntingdon Road,
Cambridge CB3 0JQ, U.K.

ABSTRACT

The disparity that exists between the physiological breeding season of the mare and the arbitrary mating period imposed upon Thoroughbred horses causes special problems in the control of anoestrus in Thoroughbred stud management. Seasonal anoestrus can now be successfully curtailed by combined therapy with artificial lighting, orally-active progestagens and prostaglandin analogues. Recent experiments indicate that low dose, slow release implants of GnRH analogues may also play a useful role in the future. Lactation anoestrus occurs mainly in mares which foal early in the year and this condition also responds well to treatment with oral progestagens. Persistence of the endometrial cup reaction following spontaneous or deliberate loss of a pregnancy after Day 40 gives rise to a unique type of gonadotrophin-induced anoestrus in the mare. This phenomenon is specially related to twinning in Thoroughbreds and is best managed by the use of ultrasound scanning for early, accurate diagnosis of pregnancy.

INTRODUCTION

The national figures for conception and foaling rates recorded annually for Thoroughbred mares in Europe and North America are in the region of only 75% and 65% respectively (General Stud Book Returns, 1972-82). A lack of selection of the breeding population for fertility potential, a determination to breed from geriatric mares due to proven genetic value and the existing international ban upon the use of artificial insemination in Thoroughbreds, all contribute significantly to these low fertility figures in the breed. But perhaps the single biggest cause of barrenness in Thoroughbred mares is the simple absence of mating that results from their failure to express ovulatory oestrous cycles at the desired time.

Three distinct types of anoestrus have practical significance in Thoroughbred stud management; those associated with season, lactation, and early pregnancy loss. This paper examines the underlying causes of anoestrus in the mare and discusses the efficacy of the various methods that are used to control the condition.

SEASONAL ANOESTRUS

The mare is a long-day, seasonally polyoestrous species in which

cyclical ovarian activity commences in early Spring, rises steadily to a
peak at the summer solstice and declines again in Autumn
(Osborne, 1966). Most mares pass into a state of deep anoestrus during
Winter although some 10-15% of Thoroughbred mares actually remain in
prolonged dioestrus due to persistence of luteal tissue from the last
ovulation in the preceeding autumn. The change from deep anoestrus to
renewed cyclical activity rarely occurs abruptly. Instead the majority of
mares pass through a variable transition phase of shallow anoestrus that
is characterized clinically by the presence of multiple non-ovulatory
follicles in the ovaries and the expression of erratic, and frequently
prolonged, oestrous behaviour.

For the purposes of age-related racing, Thoroughbred foals in the
Northern hemisphere are officially aged from January 1st irrespective of
when in the year they are actually born. Gestation lasts for $\pm$340 days in
the mare so the official mating period or "covering season" for
Thoroughbreds runs from February 15th to July 15th. The resulting
disparity between this arbitrary covering season and the mares'
physiological breeding season means that barren and maiden mares which do
not begin to cycle until mid-April effectively have as few as 2-3 oestrous
cycles in which they may be advantageously mated to the chosen stallion.
With the ban on the use of A.I. coupled with an average "book" of around
45 mares for most stallions, the cumulative effects of any prolongation of
seasonal anoestrus become obvious.

In the management system for Thoroughbreds that currently prevails in
the U.K. most non-pregnant mares arrive at the stallion studfarm during
the first fortnight in February. The winter "home environment" of these
mares varies greatly such that they can be in any one of 5 reproductive
states on February 15th. They may be: a) already cycling normally; b) in
deep anoestrus with no follicles in the ovaries; c) in shallow anoestrus
with partial follicular development; d) in prolonged Spring oestrus in the
absence of an ovulatory follicle; e) in prolonged dioestrus with peristent
luteal tissue from the previous year. Clinically it is difficult to
differentially diagnose any of these states without the aid of plasma
progesterone measurements and there thus exists a need for a therapeutic
regime that can be applied to all non-pregnant mares soon before the start
of the covering season regardless of their reproductve status at that
time.

As a long-day breeder, increasing photoperiod is the basic stimulus to renewed ovarian activity in the anoestrous mare. Burkhardt (1947) first demonstrated a sigificant hastening of renewed reproductive activity in Pony mares maintained under artificial lighting during Winter and many subsequent studies have confirmed and extended these original observations. Notably, Palmer (1979) determined that a sudden increase in light, to give a 16-hr daylength from the winter solstice, is just as effective as a step-wise increase designed to mimic the natural daylength change of Spring. More recently, the same group identified a 1-hr window, between 9.5 and 10.5 hours after dusk, when the application of artificial light is as effective as a full 16-hr daylength (Palmer, Driancourt and Ortavant, 1982).

A significant number of non-pregnant Thoroughbred mares are now subjected to a 16-hr daylength from around the time of the winter solstice. Such treatment is clearly successful in hastening the transition from deep to shallow anoestrus although considerable individual variation still occurs within any light-treated group of mares such that first ovulations of the season occur randomly between 6 and 12 weeks after the start of light therapy. This lack of precision in the response has led, in recent years, to the successful application of various progesterone-withdrawal treatments, either in conjunction with light or given alone.

Loy and Swan (1966) originally demonstrated that daily intramuscular (i/m) injection of cycling mares with 100-400 mg progesterone in oil for 15 days will suppress oestrus and ovulation to give a typical rebound oestrus that begins 2-5 days after the last injection of progesterone. More recently the same laboratory described a successful therapeutic regime for light-treated shallow anoestrous mares that involves daily i/m injections of a mixture of 150 mg progesterone and 10 mg oestradiol-17β for 10-15 days; 81% of 128 treated mares ovulated 9-16 days after the last injection (Taylor, Pemstein and Loy, 1982). However, daily intramuscular therapy has practical disadvantages in routine stud management and hence a major breakthrough was achieved with the development of the orally-active, synthetic progestagen, allyl trenbolone (Regumate; Hoechst Pharmaceuticals), for use in horses. Webel (1975) first demonstrated a suppression of oestrus and ovulation in Pony mares fed 0.44 mg/Kg allyl trenbolone daily for 15 days and Palmer (1979) then described a hastening

and synchronization of the first ovulation of the breeding season in
light-treated shallow anoestrous mares similarily admininstered allyl
trenbolone for 15 days. Subsequently Allen et al (1980) reported the
successful use of Regumate to induce a fertile oestrus and ovulation in
84% of 61 Thoroughbred mares showing seasonal or lactation anoestrus
(Figure 1) and Webel and Squires (1982) similarly described the value of
Regumate in a large clinical trial involving 477 anoestrous mares of
various breeds.

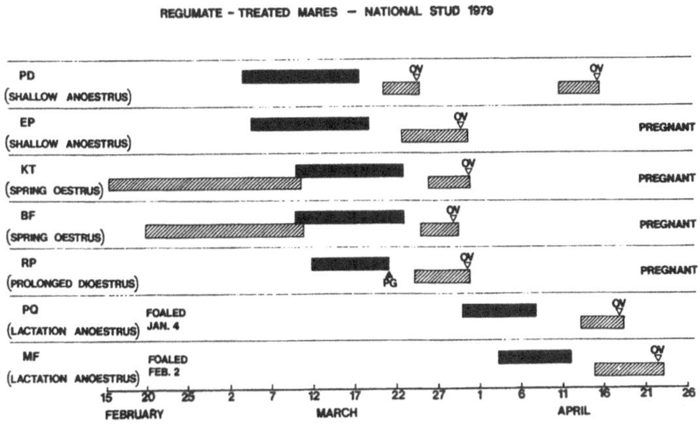

REGUMATE - TREATED MARES — NATIONAL STUD 1979

Fig. 1   Response of non-cycling Thoroughbred mares to daily
administration of allyl trenbolone (black horizontal bars)
early in the covering season.

From these and other studies it has become clear that Regumate
provides a practical and efficaceous method for inducing oestrus and
ovulation in a high proportion of mares showing shallow anoestrus or
prolonged spring oestrus.   The response is much lower in deep anoestrus,
however, so there still exists a need for pre-treatment with lights to
ensure a transition from deep to shallow anoestrus before commencing
Regumate therapy.   Furthermore, the occurrence of prolonged dioestrus in
at least 10% of mares, together with the unrecognized onset of normal
cyclicity in other mares as a result of the light treatment alone,
necessitates the administration of prostaglandin to all the mares at the
conclusion of Regumate therapy to induce luteolysis in those animals which
are already in dioestrus.

Early attempts to stimulate follicular growth in anoestrous mares
with Gonadotrophin-Releasing Hormone (GnRH) proved of little practical
value due to the need for multiple frequent injections and the
incompetence of the corpus luteum which resulted from induced ovulation
(see Evans and Irvine, 1979). Interest in GnRH therapy has been rekindled
recently, however, by the demonstration of ovulation and normal luteal
function in anoestrous ewes given a low dose constant infusion of GnRH
over several days (McLeod, Haresign and Lamming, 1983). The preliminary
results of an experiment begun this year in the author's laboratory have
likewise indicated that constant low-dose treatment with GnRH may provide
a practical method for overcoming anoestrus in the mare. Four Pony and 2
Thoroughbred mares in shallow anoestrus in early March were each given a
subcutaneous implant of a potent GnRH analogue (ICI 118630) designed to
give a constant release of 60 ug/24 hr for a period of 30 days. The 4
ponies and one of the Thoroughbreds showed strong oestrous behaviour
within 6 days after insertion of the implant and these 5 mares all
ovulated within the next 7 days; 2 of the 3 mares that were mated during
the GnRH-induced oestrus conceived. Peripheral blood FSH, LH and
progesterone assays showed a steady rise in LH levels immediately after
insertion of the implant that resulted in a typically broad peak around
the time of ovulation. Conversely, FSH concentrations began to fall
steeply within 2-3 days after treatment, presumably as a result of the
secretion of a follicular product with negative feedback activity on
pituitary FSH release (Figure 2).

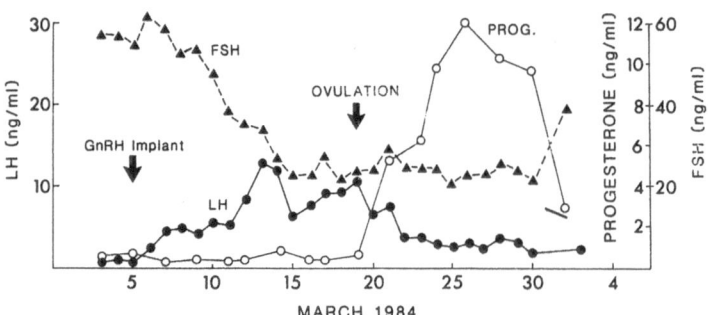

Fig. 2   Peripheral serum FSH, LH and progesterone profiles in
an anoestrous Pony mare given an implant of ICI 118630 in early
March.

103

Much basic work remains to be done to determine optimum dose rate, treatment interval and the responsiveness of mares in varying depths of anoestrus to this type of low dose, constant infusion treatment with GnRH analogues. The preliminary findings are nonetheless encouraging and it is possible to foresee a future combined therapy of artificial light, Regumate, GnRH and prostaglandin that will give a precisely timed and fertile ovulation in Thoroughbred mares on any specified day in the offical covering season.

LACTATION ANOESTRUS

Equids are unique amongst the large domestic species in their ability to express a fertile, ovulatory oestrus within as little as 9 days after parturition - the so-called foal heat. This unusual reproductive stratagem has presumably evolved to cope with the combined problems of an 11 month gestation and the ability of the mare to carry only a single foal to term. Genuine anoestrus associated with lactation occurs very rarely in those breeds of horses in which mating, and as a result foaling, lie within the confines of the physiological breeding season. But it occurs much more frequently in Thoroughbred mares and especially in animals that foal early in the year. The condition seems more closely related to photoperiod than to lactation per se since the recent studies of Palmer and his colleagues (E. Palmer, this volume) have demonstrated a significant reduction in the incidence of lactation anoestrus in early-foaling treated with a 16-hr daylength during winter.

In routine stud management many Thoroughbred mares are not mated at the foal heat due to inadequate involution of the uterus. This tendency to avoid mating so soon after foaling is greater during the earlier part of the covering season in the belief that there is still plenty of time before the need to confront the uterus with the considerable bacterial challenge of coitus. Yet it is in these early foaling mares that the chance of passing into lactation anoestrus is greatest. A further complication arises from the relatively high incidence in mares of spontaneous prolongation of luteal function in the absence of pregnancy (prolonged dioestrus; Allen and Rossdale, 1973; Stabenfeldt, Hughes, Evans and Geschwind, 1975). Thus, a sizeable proportion of foaling mares which cease to cycle are actually in prolonged dioestrus and so require only treatment with prostaglandin to destroy the persistant luteal tissue. Plasma progesterone measurement is the only valid way to accurately

104

distinguish between anoestrus and prolonged dioestrus.

As for barren and maiden in shallow anoestrus, oral administration of
Regumate has provided a good, practical treatment for mares in lactation
anoestrus (Figure 2; Allen et al, 1980). But in view of the above-
mentioned incidence of prolonged dioestrus a commonly used regime is to
first treat the non-cycling, non-pregnant lactating mare with
prostaglandin. Those animals which fail to show oestrus within 5-7 days
are then presumed to be in true anoestrus and are therefore started on the
10-day course of Regumate.

ABORTION ANOESTRUS

A third type of anoestrus which is unique to equids and which can
cause a particular problem in Thoroughbreds is that which is associated
with early pregnancy loss during the period of the equine endometrial cup
reaction between Days 40 and 120 of gestation.

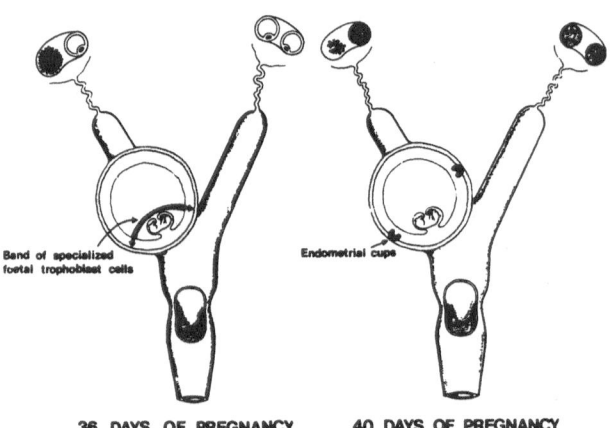

Band of specialized
foetal trophoblast cells

Endometrial cups

36 DAYS OF PREGNANCY   **40 DAYS OF PREGNANCY**

Fig. 3   Invasion of the chorionic girdle cells to form the
equine endometrial cups between Days 36 and 38 after ovulation.

The eCG-secreting endometrial cups are derived from specialized
trophoblast cells from the chorionic girdle region of the foetal membranes
which invade the maternal endometrium between Days 36 and 38 (Allen,
Hamilton and Moor, 1973). This invasion process occurs very rapidly and

the resulting cups, each consisting of a clone of gonadotrophin-secreting foetal cells within the endometrial stroma, retain no further physical connection with the rest of the foetal membranes (Figure 3). Hence, if the foetus subsequently dies or the conceptus is deliberately removed from the uterus, the endometrial cups persist through their normal lifespan and continue to secrete large quantities of eCG into the maternal bloodstream for 60-100 days. The continuing high levels of eCG then appear to suppress further ovarian activity in the affected mare and she passes into a state of very deep anoestrus. Only when the endometrial cups have finally degenerated and eCG has disappeared from the blood, does she again resume normal cyclical ovarian activity and so have a chance to be re-mated (Allen, 1970).

In Thoroughbred breeding this unique occurrence of the eCG-induced anoestrus has frequently been associated in the past with the further problem of twinning. The diffuse, non-invasive epitheliochorial architecture of the equine placenta prevents the mare from successfully carrying twin conceptuses to full term. Of those mares which do conceive twins the great majority abort between 5 and 9 months of gestation as a result of competition between the placentae leading to the death of the more disadvantaged of the two foetuses (Jeffcott and Whitwell, 1973). The rates of twin ovulation and twin conception are both higher in Thorough-breds than most other breeds and for this reason it was formerly common practice in Thoroughbred stud management to withhold from mating mares which produced twin follicles during oestrus and to deliberately induce early abortion in mares which did conceive twins. However, two major problems beset this policy. First, mares which produce twin follicles in one oestrus frequently do so in subsequent oestrous periods with the result that it usually became late in the season before they were finally mated in desperation. Second, twin conceptuses were not usually diagnosed by the standard technique of rectal palpation of the uterus until after Day 40 when the endometrial cups had already begun to develop. Two foetal sacs gives rise to two sets of endometrial cups in the uterus so that deliberate abortion of twins at this late stage leaves even higher than normal levels of eCG persisting in the bloodstream for many weeks (Allen and Rossdale, 1973).

The advent of realtime ultrasound scanning for early pregnancy diagnosis in mares has been of a major benefit to the whole problem of twinning in Thoroughbreds (Simpson et al, 1982). The technique allows accurate diagnosis of twin foetal sacs from as early as Day 18 and in the case of bicornuate twins, considerable success has been achieved since the introduction of scanning by early rupture of one of the conceptuses via manual pressure on the uterus through the rectal wall. For example, attempted rupture of one of bicornuate twins before Day 30 in 73 Thoroughbred mares during the 1982 and 1983 covering seasons, resulted in 59 (81%) on-going singleton pregnancies. This is a tremendous improvement upon the former situation where < 10% of attempts to rupture one conceptus at > Day 40 resulted in the birth of a single foal. Selective rupture of one conceptus is not possible when both foetal sacs are adjacent in the same uterine horn. Prostaglandin-induced abortion of both conceptuses before Day 36 is therefore the preferred treatment in these cases, giving a high chance of conception to re-mating in the induced oestrus.

One additional benefit of the introduction of scanning to twinning has come from an increased willingness to cover mares that develop twin follicles. The time saved by such a move is considerable and its rationale has been amply supported by the results of a recent analysis of the stud and veterinary records for 5532 oestrous cycles exhibited by 2772 Thoroughbred mares standing on studfarms in the Newmarket area of U.K. during two consecutive breeding seasons (M.W. Sanderson and W.R. Allen, unpublished data). They showed that: a) some 70% of the twins conceived occur in mares which were considered to have only a "detected" (by rectal palpation) single follicle in their ovaries at the time of mating; b) mating mares with a "detected" single follicle gives a 1.7% chance of conceiving twins and a 52% chance of conceiving a singleton pregnancy; c) mating mares with "detected" twin follicles increases the twin conception rate to only 5.5% while retaining a singleton conception rate of 54%.

CONCLUSIONS

The imposition of a brief and out-of-phase mating period upon the Thoroughbred and our failure to adequately consider fertility parameters in the selection of breeding stock has made time the single biggest enemy of the stud manager and veterinary surgeon. None of the above types of anoestrus would constitute a significant problem in Thoroughbred stud management if we followed the physiological season, provided more contact between stallions and mares and allowed natural selection to eliminate twinning genes from the breeding population. Man is certainly not the Thoroughbred mare's best friend with regard to her reproduction.

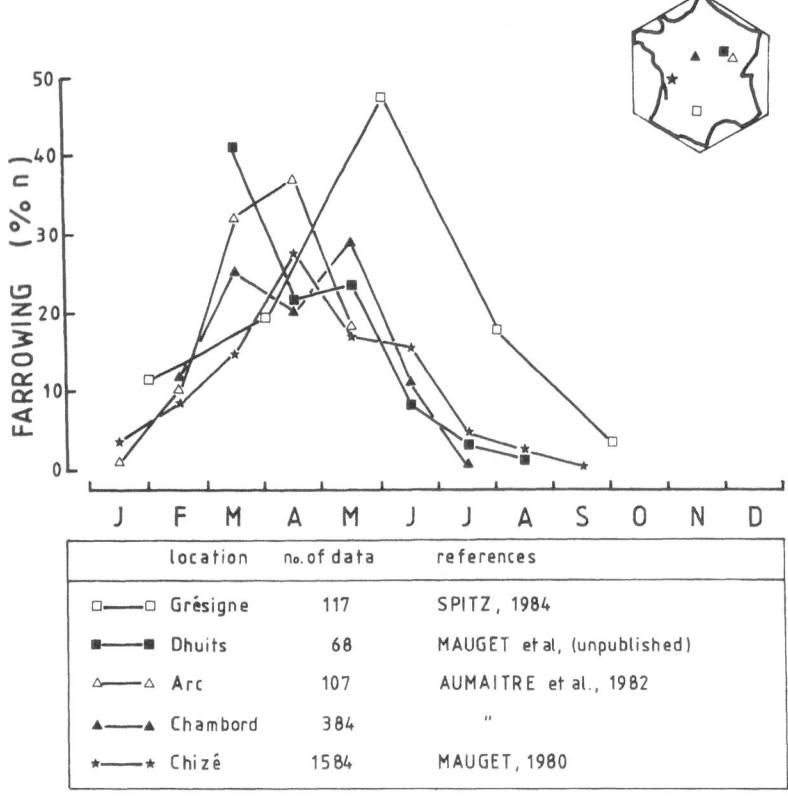

Fig. 1 Monthly distribution of farrowings in populations of European wild boar in France.

	location	no. of data	references
□——□	Grésigne	117	SPITZ, 1984
■——■	Dhuits	68	MAUGET et al, (unpublished)
△——△	Arc	107	AUMAITRE et al., 1982
▲——▲	Chambord	384	"
★——★	Chizé	1584	MAUGET, 1980

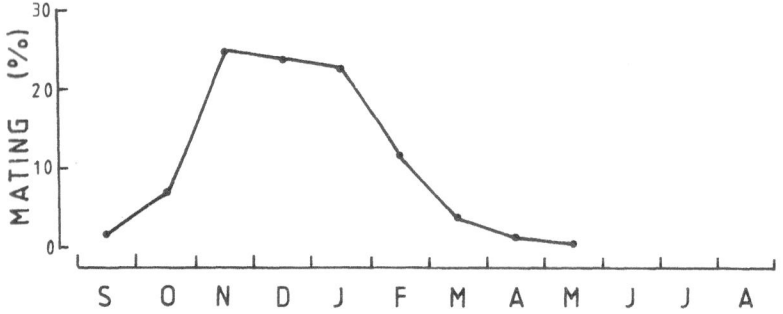

Fig. 2 Mean monthly distribution of matings (pooled data from Fig.1)

REFERENCES

Allen, W.R. (1970). Endocrinology of early pregnancy in the mare. Equine vet. J. 2, 27,32.

Allen, W.R., Hamilton, D.W. and Moor, R.M. (1973). The origin of the equine endometrial cups. II Invasion of the endometrium by trophoblast. Anat. Rec. 177, 485-501.

Allen, W.R. and Rossdale, P.D. (1973). A preliminary study on the use of prostaglandins for inducing oestrus in non-cycling Thoroughbred mares. Equine vet. J. 5, 1-4.

Allen, W.R., Urwin, V., Simpson, D.J., Greenwood, R.E.S., Crowhurst, R.C., Ellis, D.R., Ricketts, S.W., Hunt, M.D.N. and Wingfield-Digby, N.J. (1980). Preliminary studies on the use of an oral progestagen to induce oestrus and ovulation in seasonally anoestrous Thoroughbred mares. Equine vet. J. 12, 141-145.

Bunkhardt, J. (1947). Transition from anoestrus in the mare and the effects of artificial lighting. J. Agric. Sci., Camb. 37, 64-68.

Evans, M.J. and Irvine, C.H.G. (1979). Induction of follicular development and ovulation in seasonally acyclic mares using gonadotrophin-releasing hormones and progesterone. J. Reprod. Fert., Suppl. 27, 113-121.

Jeffcott, L.B. and Whitwell, K.E. (1973). Twinning as a cause of foetal and neonatal loss in Thoroughbred mares. J. comp. Path. 83, 91-96.

Loy, R.G. and Swan, S.M. (1966). Effects of exogenous progestagens on reproductive phenomena in mares. J. Anim. Sci. 25, 821-826.

McLeod, B.J., Haresign, W. and Lamming, G.E. (1983). Induction of ovulation in seasonally anoestrous ewes by continuous infusion of low doses of GnRH. J. Reprod. Fert. 68, 489-495.

Osborne, V.E. (1966). An analysis of the pattern of ovulation as it occurs in the annual reproductive cycle of the mare in Australia. Aust. vet. J. 42, 149-154.

Palmer, E. (1979). Reproductive management of mares without detection of oestrus. J. Reprod. Fert., Suppl. 27, 263-270.

Palmer, E. Driancourt, M.A. and Ortavant, R. (1982). Photoperiodic stimulation of the mare during winter anoestrus. J. Reprod. Fert., Suppl. 32, 275-282.

Simpson, D.J., Greenwood, R.E.S., Ricketts, S.W., Rossdale, P.D., Sanderson, M. and Allen, W.R. (1982). Use of ultrasound echography for the early diagnosis of single and twin pregnancy in the mare. J. Reprod. Fert., Suppl. 32, 431-439.

Stabenfeldt, G.H., Hughes, J.D., Evans, J.W. and Geschwind, I.I. (1975). Unique aspects of the reproductive cycle of the mare. J. Reprod. Fert., Suppl. 23, 155-160.

Taylor, T.B., Pemstein, R. and Loy, R.G. (1982). Control of ovulation in mares in the early breeding season with ovarian steroids and prostaglandin. J. Reprod. Fert., Suppl. 32, 219-224.

Webel, S.K. (1975). Estrus control in horses with a progestin. J. Anim. Sci. 41, 385.

Webel, S.K. and Squires, E.L. (1982). Control of the oestrous cycle in mares with altrenogest. J. Reprod. Fert., Suppl. 32, 193-198.

SEASONAL ANOESTRUS IN WILD SOWS

R. Mauget

Centre d'Etudes Biologiques des Animaux Sauvages/C.N.R.S.
Villiers-en-Bois   79360 Beauvoir-sur-Niort   France

ABSTRACT

This paper summarizes the current state of knowledge of the European wild boar reproduction. Data on reproductive performances established in several natural wild populations show a general pattern of seasonal breeding. Although the boars are reproductive all year round, the sows exhibit an annual cycle of ovarian activity, an anoestrus period, characterized by non-ovulatory levels of plasma oestrogen and basal plasma progesterone values, occurring during the summer and autumn months. A seasonal rhythm in plasma prolactin levels has been shown, following the seasonal changes in natural daylength. These results suggest that the seasonal differences in domestic sow reproduction might be a vestige of an ancestral photoperiodic rhythm.

INTRODUCTION

The domestication of the wild boar dates from the early neolithic period (Protsch and Berger, 1973) when human societies began to establish permanent settlements. It was contemporary with the domestication of other species such as sheep, goat and cow representing our present-day farm animals. The domestication process results in a modification of the gene-pool of the wild population. Artificial selection has been directed toward an improvement in growth, nutrition and reproduction. This third goal is not completely achieved yet. As discussed in this seminar, a seasonal anoestrus still persists in many breeds of farm animals nowadays.

The pig is generally known to be reproductive throughout the year. However, an increasing amount of literature reports the evidence of a marked trend to seasonal fluctuations in reproductive performances of the domestic sow. In a number of studies in different countries, summer reproductive disorders have been reported (Corteel et al., 1969; Radev et al., 1976; Paterson et al., 1978; Enne et al., 1979; Stork, 1979; Hurtgen et al., 1980; Keindorf et Plescher, 1981; Martinat-Botte et al., 1981; Dobao et al., 1983).

Since the economic interest of this fact is obvious, efforts are currently being made towards a better understanding and "solutionning" of this problem in pig reproduction.

The species Sus scrofa is a particularly interesting one in that we still have the opportunity to study the ancestral wild form : the european wild boar. From the knowledge of the reproductive biology of this wild species might result ways to investigate in the domestic counterpart.

PERFORMANCES AND SEASONALITY OF REPRODUCTION

In Table 1 are gathered, for comparison, the reproduction performance of the wild, feral and domestic forms of Sus scrofa. It clearly appears that domestication has genetically enhanced the ovulation rate. The essential difference between the forms concerns the farrowing period : whereas the domestic and feral (domesticated animals, = genetically modified, returned to a wild status) forms are reproductive throughout the year, the wild animals are seasonal breeders, farrowing occurring generally once a year.

Reproduction of european wild sow has been particularly studied in France on a number of populations in several locations. The monthly distribution of farrowing (Fig. 1) shows a marked seasonal pattern, with the maximum of births occurring in spring. These data, expressed in mean monthly percentages of mating (Fig. 2) reveal a general pattern in which mating is practically absent (< 5 %) during the summer and autumn months.

TABLE 1   Reproductive parameters in the different forms of Sus scrofa.

		Ovulation rate	Litter size	Farrowing Frequency	Period	References
European wild boar	France	5.26	4.6	1→2	jan.→ sept.	Mauget, 1972
	G.D.R.	6	5.3	1	spring-summer	Stubbe and Stubbe,19
	France	5.27	4.5	1	spring-summer	Aumaitre et al., 19
Feral Pig	U.S.A.	8.5	5.6	2	jan.→ dec.	Barrett, 1978
		8.7	6.2		-	Hagen and Kephart, 1980
Domestic Pig		15-17	12	→ 2.5	jan.→ dec.	Asdell, 1964

REFERENCES

Allen, W.R. (1970). Endocrinology of early pregnancy in the mare. Equine vet. J. 2, 27,32.

Allen, W.R., Hamilton, D.W. and Moor, R.M. (1973). The origin of the equine endometrial cups. II Invasion of the endometrium by trophoblast. Anat. Rec. 177, 485-501.

Allen, W.R. and Rossdale, P.D. (1973). A preliminary study on the use of prostaglandins for inducing oestrus in non-cycling Thoroughbred mares. Equine vet. J. 5, 1-4.

Allen, W.R., Urwin, V., Simpson, D.J., Greenwood, R.E.S., Crowhurst, R.C., Ellis, D.R., Ricketts, S.W., Hunt, M.D.N. and Wingfield-Digby, N.J. (1980). Preliminary studies on the use of an oral progestagen to induce oestrus and ovulation in seasonally anoestrous Thoroughbred mares. Equine vet. J. 12, 141-145.

Bunkhardt, J. (1947). Transition from anoestrus in the mare and the effects of artificial lighting. J. Agric. Sci., Camb. 37, 64-68.

Evans, M.J. and Irvine, C.H.G. (1979). Induction of follicular development and ovulation in seasonally acyclic mares using gonadotrophin-releasing hormones and progesterone. J. Reprod. Fert., Suppl. 27, 113-121.

Jeffcott, L.B. and Whitwell, K.E. (1973). Twinning as a cause of foetal and neonatal loss in Thoroughbred mares. J. comp. Path. 83, 91-96.

Loy, R.G. and Swan, S.M. (1966). Effects of exogenous progestagens on reproductive phenomena in mares. J. Anim. Sci. 25, 821-826.

McLeod, B.J., Haresign, W. and Lamming, G.E. (1983). Induction of ovulation in seasonally anoestrous ewes by continuous infusion of low doses of GnRH. J. Reprod. Fert. 68, 489-495.

Osborne, V.E. (1966). An analysis of the pattern of ovulation as it occurs in the annual reproductive cycle of the mare in Australia. Aust. vet. J. 42, 149-154.

Palmer, E. (1979). Reproductive management of mares without detection of oestrus. J. Reprod. Fert., Suppl. 27, 263-270.

Palmer, E. Driancourt, M.A. and Ortavant, R. (1982). Photoperiodic stimulation of the mare during winter anoestrus. J. Reprod. Fert., Suppl. 32, 275-282.

Simpson, D.J., Greenwood, R.E.S., Ricketts, S.W., Rossdale, P.D., Sanderson, M. and Allen, W.R. (1982). Use of ultrasound echography for the early diagnosis of single and twin pregnancy in the mare. J. Reprod. Fert., Suppl. 32, 431-439.

Stabenfeldt, G.H., Hughes, J.D., Evans, J.W. and Geschwind, I.I. (1975). Unique aspects of the reproductive cycle of the mare. J. Reprod. Fert., Suppl. 23, 155-160.

Taylor, T.B., Pemstein, R. and Loy, R.G. (1982). Control of ovulation in mares in the early breeding season with ovarian steroids and prostaglandin. J. Reprod. Fert., Suppl. 32, 219-224.

Webel, S.K. (1975). Estrus control in horses with a progestin. J. Anim. Sci. 41, 385.

Webel, S.K. and Squires, E.L. (1982). Control of the oestrous cycle in mares with altrenogest. J. Reprod. Fert., Suppl. 32, 193-198.

SEASONAL ANOESTRUS IN WILD SOWS

R. Mauget

Centre d'Etudes Biologiques des Animaux Sauvages/C.N.R.S.
Villiers-en-Bois   79360 Beauvoir-sur-Niort   France

ABSTRACT

This paper summarizes the current state of knowledge of the European wild boar reproduction. Data on reproductive performances established in several natural wild populations show a general pattern of seasonal bree- ding. Although the boars are reproductive all year round, the sows exhibit an annual cycle of ovarian activity, an anoestrus period, characterized by non-ovulatory levels of plasma oestrogen and basal plasma progesterone va- lues, occurring during the summer and autumn months. A seasonal rhythm in plasma prolactin levels has been shown, following the seasonal changes in natural daylength. These results suggest that the seasonal differences in domestic sow reproduction might be a vestige of an ancestral photoperiodic rhythm.

INTRODUCTION

The domestication of the wild boar dates from the early neolithic pe- riod (Protsch and Berger, 1973) when human societies began to establish permanent settlements. It was contemporary with the domestication of other species such as sheep, goat and cow representing our present-day farm ani- mals. The domestication process results in a modification of the gene-pool of the wild population. Artificial selection has been directed toward an improvement in growth, nutrition and reproduction. This third goal is not completely achieved yet. As discussed in this seminar, a seasonal anoes- trus still persists in many breeds of farm animals nowadays.

The pig is generally known to be reproductive throughout the year. Ho- wever, an increasing amount of literature reports the evidence of a marked trend to seasonal fluctuations in reproductive performances of the domestic sow. In a number of studies in different countries, summer reproductive di- sorders have been reported (Corteel et al., 1969; Radev et al., 1976; Pa- terson et al., 1978; Enne et al., 1979; Stork, 1979; Hurtgen et al., 1980; Keindorf et Plescher, 1981; Martinat-Botte et al., 1981; Dobao et al., 1983).

Since the economic interest of this fact is obvious, efforts are cur- rently being made towards a better understanding and "solutionning" of this problem in pig reproduction.

The species Sus scrofa is a particularly interesting one in that we still have the opportunity to study the ancestral wild form : the european wild boar. From the knowledge of the reproductive biology of this wild species might result ways to investigate in the domestic counterpart.

PERFORMANCES AND SEASONALITY OF REPRODUCTION

In Table 1 are gathered, for comparison, the reproduction performances of the wild, feral and domestic forms of Sus scrofa. It clearly appears that domestication has genetically enhanced the ovulation rate. The essential difference between the forms concerns the farrowing period : whereas the domestic and feral (domesticated animals, = genetically modified, returned to a wild status) forms are reproductive throughout the year, the wild animals are seasonal breeders, farrowing occurring generally once a year.

Reproduction of european wild sow has been particularly studied in France on a number of populations in several locations. The monthly distribution of farrowing (Fig. 1) shows a marked seasonal pattern, with the maximum of births occurring in spring. These data, expressed in mean monthly percentages of mating (Fig. 2) reveal a general pattern in which mating is practically absent (< 5 %) during the summer and autumn months.

TABLE 1   Reproductive parameters in the different forms of Sus scrofa.

		Ovulation rate	Litter size	Farrowing Frequency	Farrowing Period	References
European wild boar	France	5.26	4.6	1→2	jan.→ sept.	Mauget, 1972
	G.D.R.	6	5.3	1	spring-summer	Stubbe and Stubbe,1977
	France	5.27	4.5	1	spring-summer	Aumaitre et al., 1982
Feral Pig	U.S.A.	8.5	5.6	2	jan.→ dec.	Barrett, 1978
		8.7	6.2		-	Hagen and Kephart, 1980
Domestic Pig		15-17	12	→ 2.5	jan.→ dec.	Asdell, 1964

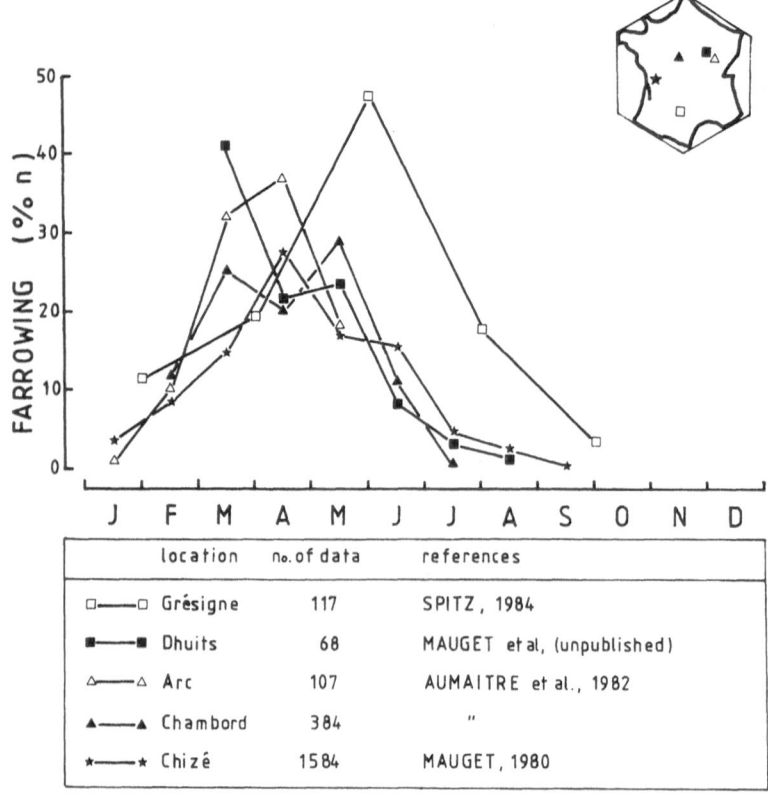

Fig. 1   Monthly distribution of farrowings in populations of European
wild boar in France.

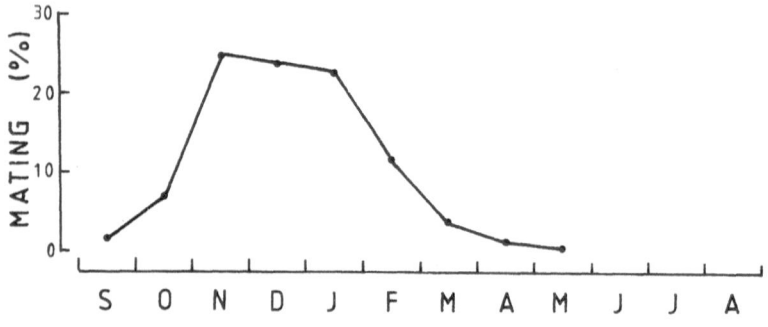

Fig. 2   Mean monthly distribution of matings (pooled data from Fig.1)

ANNUAL CYCLE OF GONADAL ACTIVITY

The seasonality of reproduction deduced from the above data on wild populations necessarly implies the existence of a sexually inactive period.

Male sexual activity

Sexual maturity occurs around 10 months (30-35 kg) of age (Henry, 1966; Duncan, 1979; Mauget, 1980). Once puberty is attained, the mean plasma androgen level remains significantly higher than in immature animals (1.46 ± 0.33 vs 0.39 ± 0.04 ng/ml, respectively). However, like numerous wild male mammals, the wild boar exhibits seasonal changes in the endocrine testis function (Fig. 3). The maximal testosterone levels are observed in late autumn and winter. But the exocrine testis function remains active throughout the year (Mauget, 1980). Thus the possibility of mating theoretically exists.

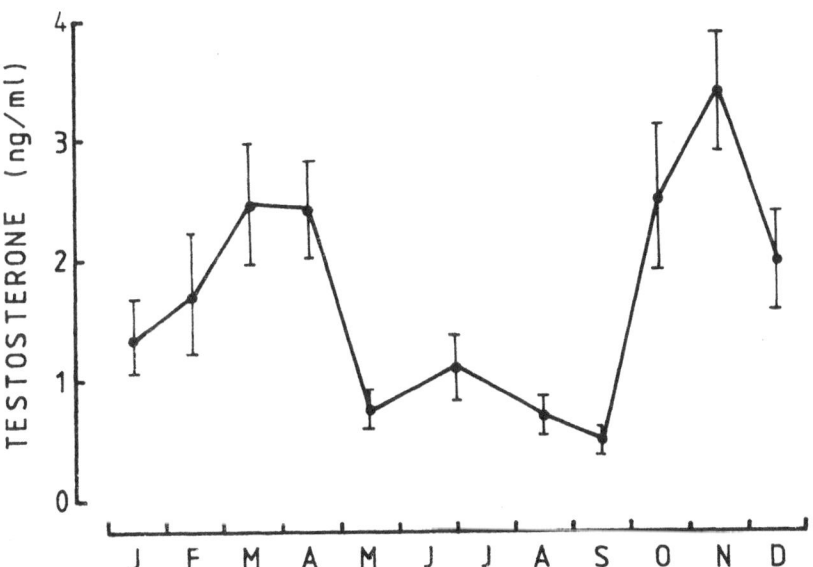

Fig. 3 Annual cycle of plasma testosterone in the wild boar (monthly mean values ± s.e.m. calculated over a period of two years).

Female sexual activity

Conversely to the male, a wide range in age (7 to 22 months) and weight (35 to 70 kg) at puberty is observed in the wild sow (Mauget, 1980; Aumaitre et al., 1982). Irrespective of the season of birth, the attainment of female puberty shows a seasonal variation, as puberty only occurs from

October to June (Fig. 4). It seems that, once a critical growth stage is reached, puberty might or might not occur according to the time of the year. This must be further investigated and compared with seasonal differences in LH production reported in the prepuberal domestic sow (Foxcroft et al., 1979).

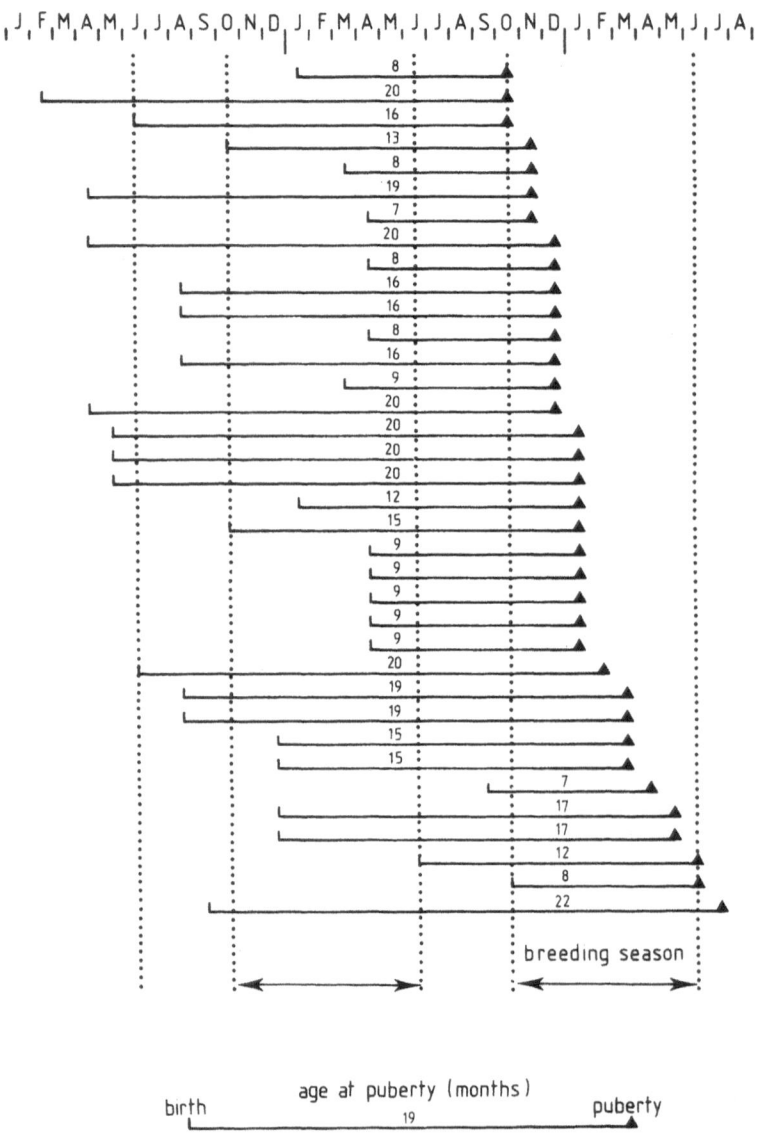

Fig. 4   Range of season and age the attainment of puberty (individual data from a commercial rearing unit).

Data from adult females captured in the wild (Chizé Forest, Midwes-
tern France) led us to establish the distribution of reproductive states
according to the time of year and the plasma progesterone levels related to
these states (Fig. 5). The interesting point to retain is the occurrence of
anoestrus animals during the summer and autumn months. In the females with
no functional corpora lutea (immature, lactating or females seasonally in
anoestrus) a stress induced release of progesterone by the adrenal cortex
contributes to increasing the basal level.

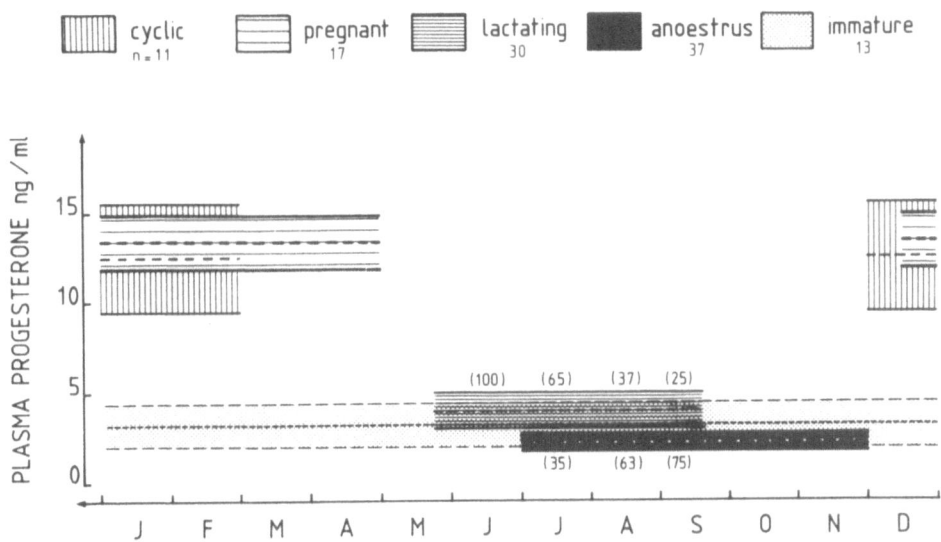

Fig. 5   Distribution of reproductive state on 108 (=Σn) female wild
boars captured in the wild and related plasma progesterone levels
(mean ± 2 s.e.m; : 9.95 confidence limits).
( ) : relative frequency, in %, of lactating and anoestrous states.

In unbred adult females (kept in outdoor enclosures without males) se-
rially sampled (on a weekly basis), seasonal variations in ovarian activity
appear through characterization of key ovarian steroids, namely progesterone and estradiol-17 β. Figure 6 summarizes this evolution. There is a regular succession of cyclic ovarian activity (winter-spring) and anoestrus sequences (summer-autumn).

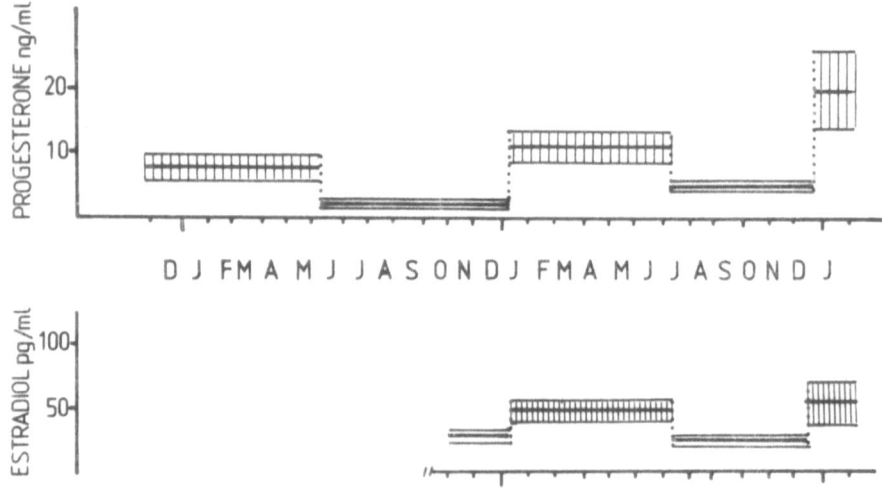

Fig. 6   Seasonal variations in progesterone and estradiol-17 β mean levels (mean ± 2 s.e.m. ∴ 0.95 confidence limits) in unbred adult females (pooled data from weekly measures).

FACTORS INVOLVED IN ANOESTRUS CONTROL

In the seasonal breeders, the synchronization of sexual activity might result from the adjustment of an "endogenous oscillatory system" (Assenmacher, 1979) by cyclic environmental factors. In the wild boar two main factors could be involved : nutritional factors and photoperiodism.

In rearing, under stable optimum feeding conditions, there always appears an anoestrus period whose minimum duration extends from July to September (Mauget, 1978). In the forest, nutritional conditions, particularly the autumnal acorn production, fluctuate on a yearly basis. It has been established that earliness or delay in the onset of the breeding season (October or January) is related to a respectively high or low natural food availability. Furthermore, increased food results in a higher average ovulation rate (Aumaitre et al., 1984).

For numerous species, natural photoperiodism is the main synchronizer of seasonal sexual activity (Menaker et al., 1978). The sensitivity of the wild boar to the seasonal changes in daylight length can be assessed from the annual variation in plasma prolactin concentration. Figure 7 shows the seasonal rhythm of prolactin secretion established in wild sows. A significative correlation ($r = 0.9$, $P < 0.001$) exists between the level of prolactin and the duration of day light (Ravault et al., 1982). A similar photoperiod-prolactin correlation has been reported both in domestic animals (cattle : Karg et Schams, 1974; Leining et al., 1979; goat : Buttle, 1974; Hart, 1975; sheep : Pelletier, 1973; Lamming et al., 1979; Ravault, 1976) and in wild animals (white-tailed deer : Mirarchi et al., 1978; badger, fox : Maurel, 1981; roe deer : Semperé, 1982; mink : Boissin-Agasse et al., 1983).

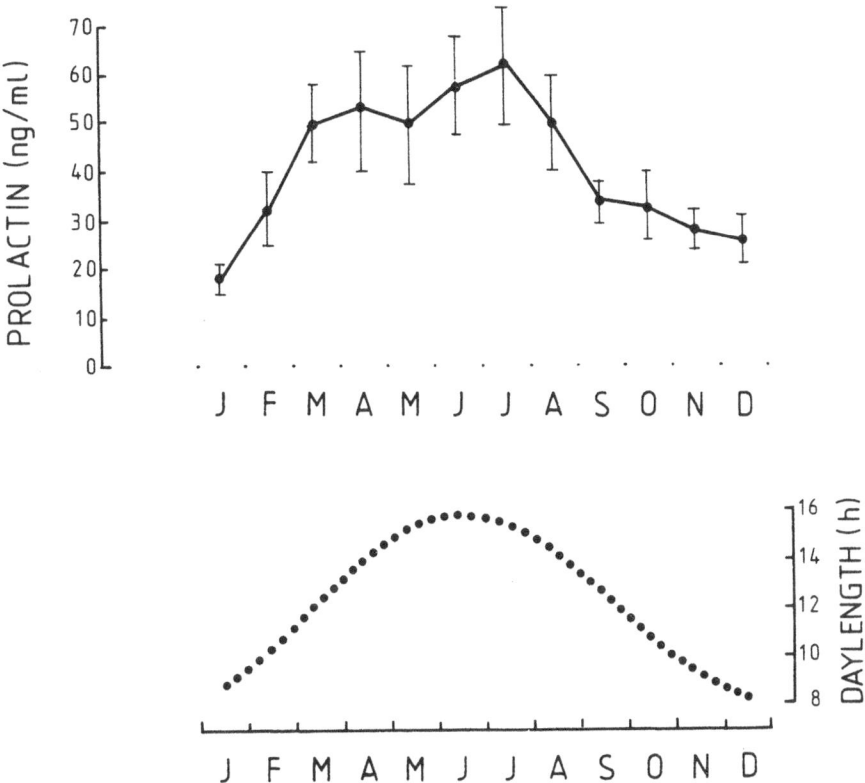

Fig. 7 - Seasonal variations in plasma prolactin levels (monthly pooled data of weekly samples obtained from 16 unbred adult females).

The data on reproductive activity in the wild boar summarized in this paper, emphasize environmental control. As pointed out recently by Hansel and Convey (1983) the concept that seasonal effects control reproductive processes in the Pig is now emerging. Among possible seasonal factors, temperature does not seem to be the prevailing one (Hurtgen et al., 1980; Dobao et al., 1983). Thus a photoperiodic involvment could be considered.

116

Ravault et al. (1982) have recently reported a seasonal influence of daylight on prolactin secretion in domestic pigs. Other reports are however
conflicting (Hoagland et al., 1981; Kraeling et al., 1983). Nevertheless,
one might consider that, at least in some genetic lines, the observed seasonal variation in reproductive activity could be related to a vestige of
an ancestral photoperiodic rhythm. However, it is now important to find out
if prolactin is directly responsible for seasonal anoestrus as from the literature this has never been clearly demonstrated. Prolactin patterns in
sows with prolonged periods of ovarian inactivity are reported to be similar to those of normal sows (Benjaminsen, 1981). Since in many seasonal
breeders the annual cycle of prolactinaemia has been found to follow the
same cyclic pattern, whatever the timing of reproduction (long or short
days), one could suggest that seasonal sexual inactivity might primarily
result from hypothalamic disorders. The feedback mechanisms between gonadal hormones and gonadotrophin could be seasonally modified by the photoperiod (Tureck and Campbell, 1979; Karsch et al., 1980). It is however unlikely that the control of sexual rhythmicity may be merely exogenous. An endogenous cyclicity in the sensitivity of the hypothalamohypophysial system
could be involved. In addition, one must consider that, at least in wild
mammals in which the endocrine adaptive mechanisms follow a distinct seasonal pattern, the sexual cycle is integrated in a multihormonal interactive
system involving, among others, the thyroid and adrenal glands.

REFERENCES

Asdell, S.A. 1964. Patterns of Mammalian Reproduction. Cornell University
    Press, New-York.
Assenmacher,I. 1974. External and internal components of the mechanism con-
    trolling reproductive cycles in Drakes. Circannual clocks. E.T. Pin-
    gelley, Ed. Acad. Press, N.-Y., 197-239.
Aumaitre, A., Morvan, C., Quere, J.P., Peiniau, J. and Valet, G. 1982. Pro-
    ductivité potentielle et reproduction hivernale chez la laie (Sus scro-
    fa scrofa) en milieu sauvage. Journées Rech. Porcine en France, 1982,
    14, 109-124.
Aumaitre, A., Quere, J.P. and Peiniau, J. 1984. Influence du milieu sur la
    reproduction hivernale et la prolificité de la laie. Symp. Int. San-
    glier. Ed. INRA Publ. 22, 69-78.
Barrett, R. 1978. The feral hog on the Dye Creek Ranch, California. Hilgar-
    dia, 46 (9), 283-355. Agric. Sc. Publ. Univ. California.

Benjaminsen, E. 1981. Plasma prolactin in the sow with emphasis on variation in resumption of ovarian activity after weaning. Acta vet. scand., 22, 67-77.

Boissin-Agasse, L., Ravault, J.P. and Boissin, J. 1983. Photosensibilité circadienne et contrôle hypophysaire du cycle annuel de la prolactiné- mie chez le Vison. C.R. Acad. Sci. (Paris). 296, 707-710.

Buttle, H.L. 1974. Seasonal variation of prolactin of male goats. J. Reprod. Fert., 37, 95.

Corteel, J.M., Signoret, J.P. and du Mesnil du Buisson, F. 1964. Variations saisonnières de la reproduction de la Truie et facteurs favorisant l'anoestrus temporel. 5th Congr. Int. Reprod. Anim. Insem. Artif., Trente 3, pp 536-546.

Dobao, M.T., Rodriganez, J. and Silio, L. 1983. Seasonal influence on fe- cundity and litter performance characteristics in Iberian pigs. Livest. Prod. Sci., 10, 601-610.

Duncan, R.W. 1974. Reproductive biology of the european wild hog in the Great Smoky Mountains National Park. M.S. Thesis, Univ. Tennessee, 95 p.

Enne, G., Beccaro, P.V. and Tarocco, C. 1979. A note on the effect of cli- mate on fertility in pigs in the Padana valley of Italy. Animal Pro- duction, 28 (1), 115-117.

Foxcroft, G.R., Stickney, K. and Edwards, S. 1979. The potential of using oestrogen for ovulatory control in the gilt and sow. Proc. Ann. Conf. Soc. Study Fertil., Glasgow. p. 18.

Hagen, D.R. and Kephart, K.B. 1980. Reproduction in domestic and feral swi- ne. I. Comparison of ovulatory rate and litter size. Biol. Reprod., 22 (3), 550-552.

Hansel, W. and Convey, E.M. 1983. Physiology of the estrous cycle. J. Anim. Sci., 57, suppl. 2, 404-424.

Henry, V.G. 1966. European wild hog hunting season recommandations based on reproductive data. Proc. Ann. Conf. S.E. Assoc. Game and Fish Comm., 20, 139-145.

Hoagland, T.A., Diekman, M.A. and Malven, P.V. 1981. Failure of stress and supplemental lighting to affect release of prolactin in swine. J. Anim. Sci., 53 (2), 467-472.

Hurtgen, J.P., Leman, A.D. and Crabo, B. 1980. Seasonal influence on es- trous activity in sows and gilts. J. American Vet. Med. Assoc., 176 (2), 119-123.

Karg, H. and Schams, D. 1974. Prolactin release in cattle. J. Reprod. Fert., 39, 463-472.

Karsch, F.J., Goodman, R.L. and Legan, S.J. 1980. Feedback basis of seaso- nal breeding : test of an hypothesis. J. Reprod. Fert., 58, 521-535.

Keindorf, A. and Plescher, W. 1981. Der Jahreszeiteneinfluss auf die Fruch- tbarkeitsleistungen der Schweine unter besonderer Berücksichtigung der Sommermonate. Mh. Vet.-Med., 36, 324-330.

Kraeling, R.R., Rampacek, G.B., Mabry, J.W., Cunningham, F.L. and Pinkert, C.A. 1983. Serum concentrations of pituitary and adrenal hormones in female pigs exposed to two photoperiods. J. Anim. Sci., 57 (5), 1243- 1250.

Lamming, G.E., Moseley, S.R. and McNeilly, J.R. 1974. Prolactin release in the sheep. J. Reprod. Fert., 40, 151-168.

Leining, K.B., Bourne, R.A. and Tucker, H.A. 1979. Prolactin response to duration and wavelength of light in prepubertal bulls. Endocrinology, 104, 289-294.

Martinat-Botte, F., Dando, P., Gautier, J. and Terqui, M. 1981. Variation saisonnière de la taille de la portée en relation avec celles du taux d'ovulation et de la mortalité embryonnaire chez la Truie. Journées Rech. Porcine en France, 1981, 269-276, I.T.P. Ed. Paris.

Mauget, R. 1972. Observations sur la reproduction du Sanglier (Sus scrofa L.) à l'état sauvage. Annls Biol. anim. Biochim. Biophys., 12 (2), 195-202.

Mauget, R. 1978. Seasonal reproductive activity of the european wild boar; comparison with the domestic sow. In : "Environmental Endocrinology", Eds. I. Assenmacher et D.S. Farner, Springer Verlag, pp. 79-80.

Mauget, R. 1980. Régulations écologiques, comportementales et physiologiques (reproduction) de l'adaptation du Sanglier, Sus scrofa, au milieu. Doct. Thesis, Univ. Tours, 297 pp.

Maurel, D. 1981. Variations saisonnières des fonctions testiculaire et thyroïdienne en relation avec l'utilisation de l'espace et du temps chez le Blaireau européen (Meles meles) et le Renard roux (Vulpes vulpes). Doct. Thesis, Univ. Montpellier, 302 pp.

Menaker, M., Takahashi, J.S. and Eskin, A. 1978 . The physiology of circadian pace makers. Ann. Rev. Physiol., 40, 501-526.

Mirarchi, R.E., Howlands, B.E., Scanlon, P.F., Kirkpatrick, R.L. and Sanford, L.M. 1978. Seasonal variation in LH, FSH, prolactin, and testosterone concentrations in adult male white-tailed deer. Can. J. Zool., 56, 121-127.

Paterson, A.M., Barker, I. and Lindsay, D.R. 1978. Summer infertility in pigs : its incidence and characteristics in an Australian commercial piggery. Australian J. of Exp. Agric. and Anim. Husbandry, 18 (94), 698-701.

Pelletier, J. 1973. Evidence for photoperiodic control of prolactin release in rams. J. Reprod. Fert., 35, 143-147.

Protsch, R. and Berger, R. 1973. Earliest radiocarbon dates for domesticated animals. Science, 179 (4070), 235-239.

Radev, G., Andreev, A. and Kostov, L. 1976. The influence of age and season on the weaning to oestrus period in sows. Intern. Congr. Anim. Reprod. Artif. Insem, Krakow, July 1976, Proceedings, vol. 1. Communication abstracts, 208.

Ravault, J.P., Martinat-Botte, F., Mauget, R., Martinat, N., Locatelli, A. and Bariteau, F. 1982. Influence of the duration of daylight on prolactin secretion in the Pig : hourly rhythm in ovariectomized females, monthly variation in domestic (male and female) and wild strains during the year. Biol. Reprod., 27, 1084-1089.

Semperé, A. 1982. Fonction de reproduction et caractères sexuels secondaires chez le Chevreuil. Doct. Thesis, Univ. Tours, 291 pp.

Spitz, F. 1984. Démographie du Sanglier en Grésigne (Sud-Ouest de la France) Symp. Int. Sanglier. Ed. INRA Publ. 22, 151-157.

Stork, M.G., 1979. Seasonal reproductive inefficiency in large pig breeding units in Britain. Vet. Rec., 104, 49-52.

Stubbe, W. and Stubbe, M. 1977. Vergleichende Beiträge zur Reproduktions- und Geburtsbiologie von wild-und Hausschwein. Beitr. Jagd-und wilforschung, X, 153-179.

Turek, F.W. and Campbell 1979. Photoperiodic regulation of neuroendocrine-gonadal activity. Biol. Reprod., 20, 32-50.

PHOTOPERIOD AND FERTILITY IN THE PIG

R. Claus*, U. Weiler*, R. Hahn**

Universität Hohenheim
*Institut für Tierhaltung und Tierzüchtung
Fachgebiet Tierhaltung - 470 -
D - 7000 Stuttgart 70, Postfach 70 05 62
**Besamungsverein Neustadt a.d. Aisch e.V.
D - 8530 Neustadt a.d. Aisch

ABSTRACT

In moderate climates a seasonal pattern of reproductive functions exists for both boars and sows. In the autumn/winter period all criteria are optimized, a depression occurs in February/March and after a transient increase (around May) mainly in July/August. In boars this pattern is observed for testicular steroid production, libido, sperm count and other criteria. In the sow a similar pattern is obvious for conception rate, the weaning-oestrus-interval and litter size.

By the application of light programmes it was found that the photoperiod is the main environmental stimulus. Consequently light programmes with decreasing daylength may be used to improve reproduction during the summer period. This was shown e. g. for male reproductive functions and for the weaning to oestrus interval in the sow.

INTRODUCTION

Although the domestic pig reproduces throughout the year, seasonal variations in reproductive criteria of both boars and sows seem to exist. Fertility tends to be optimal in autumn and winter; problems arise in summer. Thus the pattern is similar to the wild hog (Mauget, 1982).

The environmental influence which is responsible for the seasonal variation may be the high ambient temperature which is associated with summer. In tropical and subtropical climates heat stress can be regarded as the reason for a depression in fertility of boars (Steinbach, 1976) and sows (Love, 1978). Experiments in heat chambers confirm such a negative influence of high temperatures on reproduction of pigs (Wetteman and Desjardins, 1979; Cameron and Blackshaw, 1980).

In moderate climates the seasonal changes of the photoperiod have to be regarded as the regulating factor (Ortavant et al., 1964) as it was shown for several species by light programmes.

The following paper briefly reviews seasonal variations of reproductive functions of boars and sows including experiments on the influence of light programmes.

PHOTOPERIOD AND FERTILITY OF THE BOAR

Seasonal changes of reproductive criteria of the boar

Seasonal differences of several reproductive criteria of boars have been described. In subtropical Australia a difference of 13 % was found for sperm concentration when comparing values in summer and winter. Sperm counts were significantly (p < 0,05) elevated in winter (Cameron, 1980). In a large population of boars in Germany semen characteristics were evaluated in the periods of September/February and March/August. In the winter period, the percentage of motile sperms, the sperm density and consequently the total number of sperms were significantly improved (Peter et al., 1981). When analysing the 2-year records of a german AI-station a clear seasonal change of ejaculate volumes was found with minimum values in February/March and (after a transient increase in May/June) again in July (Schindler, 1980). The highest volumes were apparent in November. This resembles the annual pattern of sperm production described by Jaussiaux, (1964) with a maximum in November, a decrease in February, followed by a moderate increase in April/May and a nadir in June.

When we analysed several reproductive criteria of 4 mature boars, used for AI, in weekly intervals for more than one year (details see: Claus et al., 1983; Claus et al., 1984 a, b) we found patterns e. g. for testosterone in seminal plasma, sperms per ejaculate and the libido which are comparable to the annual patterns described above. Examples for one boar are shown in fig. 1 - 3. In the case of the total number of sperms the delay, compared to the testosterone pattern, is explained by the duration of spermatogenesis (Swierstra, 1968).

Influences of artificial light programmes on reproductive criteria of boars

Only few experiments deal with the influence of light programmes on reproductive criteria of mature boars. When comparing the influence of a constant long day (16 h light daily) with a constant short day (10 h light daily) Mazzari et al. (1968) found elevated volumes but decreased sperm output in long-day-boars.

In our study described above (Claus et al., 1983; Claus et al., 1984 a, b) four other boars, were subject to a light-reverse-programme (shortest day, 8 h, 21st July; longest day, 17 h, 21st December). This light programme also reversed the seasonal pattern of reproductive functions in the boars. Examples are shown in fig. 4 - 6. These studies show, that light programmes may improve sexual functions in boars during the summer period, the criteria investigated are significantly

ng testosterone/ml seminal plasma

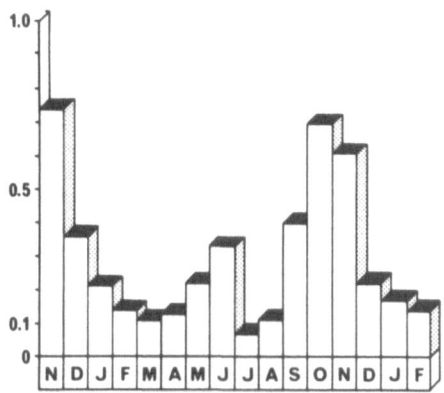

Fig. 1   Annual pattern of testosterone concentrations in seminal plasma of a boar as influenced by the
natural photoperiod (Claus et. al.,
1983, 1984 a)

reaction time (min)

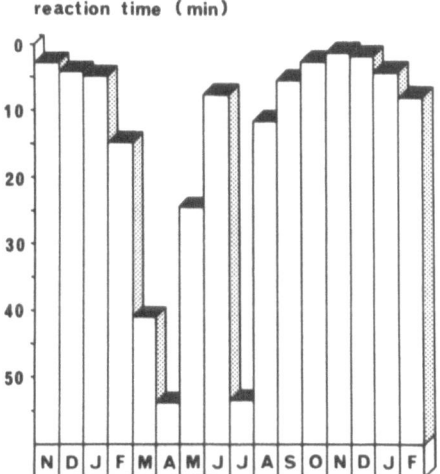

Fig. 2   Annual pattern of reaction
time (indicating libido) of a boar
as influenced by the natural photoperiod (Claus et al., 1984 b)

x 10⁹ sperms / ejaculate

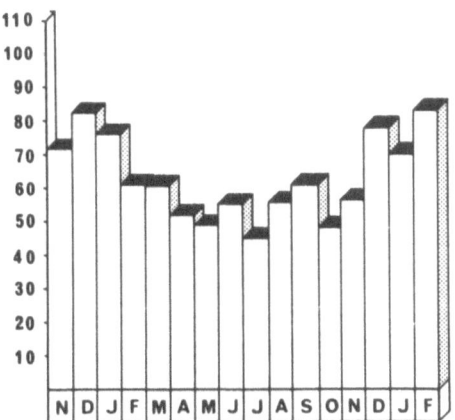

Fig. 3   Annual pattern of sperm-
counts in a boar as influenced by
the natural photoperiod. (Claus et
al., 1984 b)

122

ng testosterone/ml seminal plasma

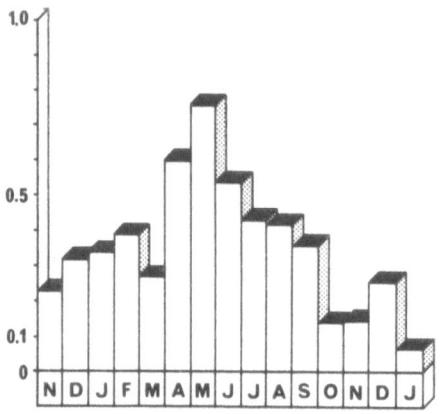

Fig. 4 Annual pattern of testosterone concentrations in seminal plasma of a boar as influenced by a
light-reverse-programme (Claus et
al., 1983, 1984 a)

reaction time (min)

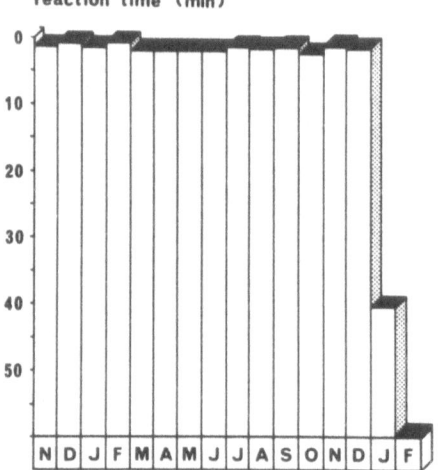

Fig. 5 Annual pattern of reaction
time (indicating libido) of a boar
as influenced by a light-reverse-
programme (Claus et al., 1984 b)

x 10⁹ sperms / ejaculate

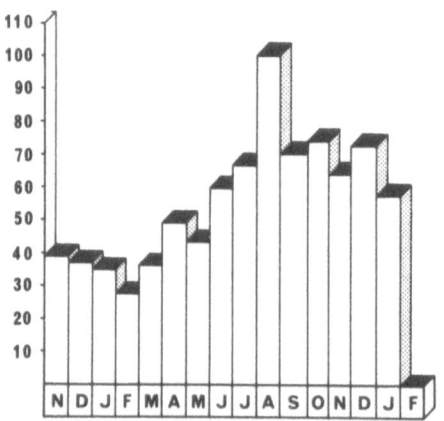

Fig. 6 Annual pattern of sperm
counts in a boar as influenced by
a light-reverse-programme (Claus
et al., 1984 b)

different during this season in favour of the light programme boars (Claus et al., 1984 a, b). Further practical consequences are currently investigated by the application of the same programme in a large AI-station and tend to confirm the results.

"Reversing" half of the boars might be of economical relevance, to have fully active boars available when others - subject to the natural photoperiod - show a depression in sexual functions. The long-term intention is, to influence animals only during the period of increasing and long-day photoperiod and maintain the beneficial effects of decreasing natural photoperiod. In a first study we therefore investigated the degree and duration of a return to the natural photoperiod in boars which were previously subject to a light reverse programme.

Influence of a return from a light reverse programme to the natural photoperiod

60 AI boars were subject to the natural photoperiod, another 52 boars were subject to a light-reverse-programme. In the beginning of November the light-programme boars returned from 14 h 40 min light per day to the natural daylength of 9 h 20 min. Fig. 7 shows the resulting testosterone concentrations in semen. In daylight boars with maximum steroid production in October (see also fig. 1) the steroid concentrations slowly decrease. Within 4 weeks the return to the normal photoperiod in the light-programme boars leads to testosteron concentrations, which are no longer statistically different from the "daylight boars".

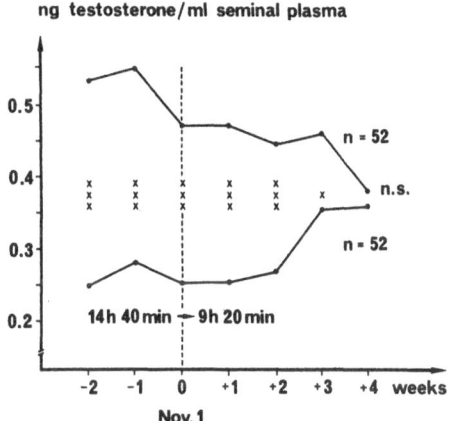

Fig. 7 Differences in seminal testosterone concentrations of AI-boars. Upper line: boars with natural photoperiod. Lower line: boars returning (week 0) from a light reverse programme to the natural photoperiod. (***: p < 0.001; *: p < 0.5) (Claß, 1984)

PHOTOPERIOD AND FERTILITY OF THE SOW

Compared to the boar, seasonal variations of reproductive criteria are still more pronounced in the sow and are often referred to as "summer infertility". A decreased reproductive efficiency of sows is observed in hot climates (e.g. Love, 1978; Paterson et al., 1978; Williamson et al., 1980) as well as in moderate

124

climates (e. g. Stork, 1979; Keindorf and Plescher, 1981; Hurtgen and Leman, 1981 a, b). Reduced fertility during the summer period may be the consequence of a variety of reproductive phenomena discussed below.

Weaning to oestrus interval as influenced by the natural photoperiod or a light-programme

After weaning the sows usually return to oestrus within one week. This interval, however, seems to be influenced by the season. Prolonged intervals are regularly reported for the summer period (Aumaitre et al., 1976; Tomes and Nielsen, 1979; Keindorf and Plescher, 1981). For example in a study involving more than 18 000 sows the percentage of sows returning to oestrus within one week was decreased in July - September (68.6 %) compared to the rest of the year (82 %) (Hurtgen et al., 1980). Additionally the abnormally delayed (> 45 days) return to oestrus was found to be two-fold during this period compared to other months (Corteel et al. 1964). During the summer period the percentage of inactive ovaries also increases (fig. 8) as found at least when investigating slaughterhouse material (Hurtgen and Leman, 1978).

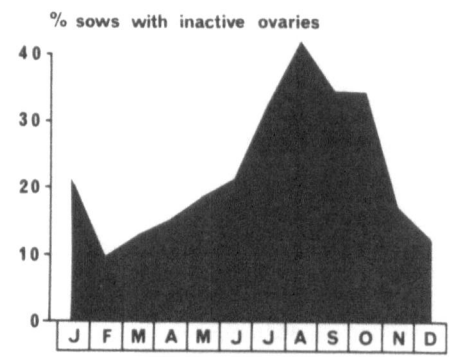

Fig. 8   Annual pattern of the percentage of sows with inactive ovaries (data from Hurtgen and Leman, 1978).

In a private farm we observed more than 110 weaning-oestrus intervals over a one year period (fig. 9, top). During the long days (see also fig. 10, top) of the natural photoperiod the mean interval from weaning to oestrus increased up to more than 30 days when weaned in August. Shortest intervals were obvious in November (4.9 days) and December (6.3 days).

In the following year, starting in May, a light programme was used, which decreased the daily light period for 20 min/week (fig. 10, lower part) until the end of August. The resulting intervals (fig. 9, lower part) decreased to an average 5.7 days in June - August. The return to the normal photoperiod (from 10 h to 13 h 31 min light/day) increased the interval in September (20 days) followed by a continuous decrease as natural daylength decreased. We suppose that a return

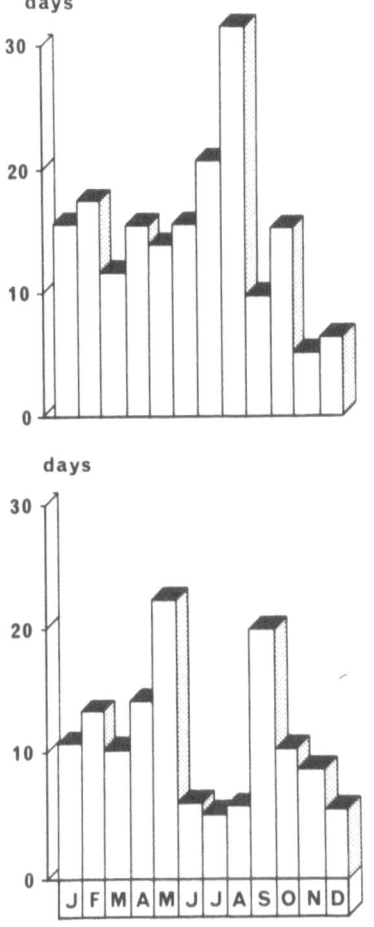

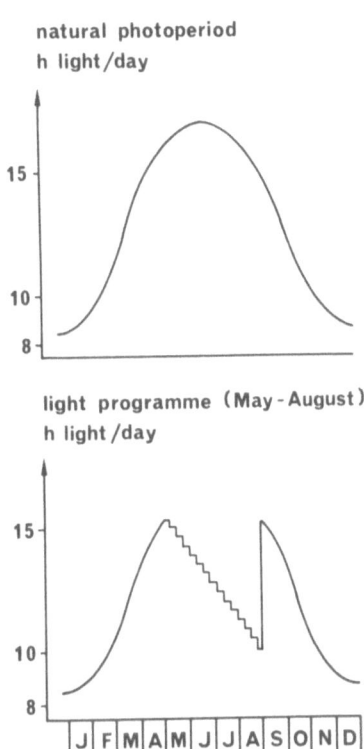

Fig. 9   Annual pattern of the weaning to oestrus interval. Upper part:
Influence of the natural photoperiod
(see: upper part, fig. 10). Lower
part: Influence of a lightprogramme
(see: lower part, fig. 10) (Claus et
al., 1984 c)

Fig. 10   Duration of the daily
light period. Upper part: natural
photoperiod. Lower part: light programme for influencing the weaning
to oestrus interval (Claus et al.,
1984 c)

to the natural photoperiod one month later might have avoided the transient
increase of the weaning/oestrus interval in September. Similarly, the results shown
in fig. 7 indicate that it seems to be possible to improve sexual functions only
during periods with lowered fertility by returning in time to the stimulating influence
of the natural photoperiod in autumn.

Seasonal pattern of the conception rate

An example for the annual pattern of conception rates is shown in fig. 11 (Corteel et al., 1964). In more recent studies similar patterns were reported (even if the absolute variations were less pronounced) and conception problems in summer are regularly emphasised for many countries (e. g. Herak, 1974; Roller and Stombaugh, 1974; Wrathall, 1977; Tomes and Nielsen, 1979; Bevier and Backstrom, (1981).

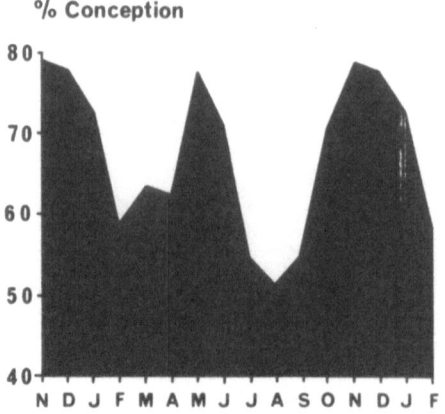

% Conception

Fig. 11   Annual pattern of the conception rate in sows (data from Corteel et al., 1964)

The decreased conception rate in summer may be explained by several influences. During summer the duration of oestrus seems to be prolonged (Signoret, 1967), so that timing of AJ may be less accurate. Additionally an increase of embryonic death following service in summer may help to explain the differences (Moore et al., 1971, Martinat-Botte et. al., 1981). Apparently the decreased conception rate is also linked to the weaning-oestrus-interval. Sows returning in summer to oestrus within one week also have an elevated conception (Van der Heyde et al., 1974). Even if variations in the conception rate are mainly related to variations of female reproductive functions, an additional effect of the boars must be considered:

In a recent study involving more than 60 000 inseminations using extended semen from boars which were kept under natural photoperiod and another more than 60 000 inseminations from boars with a light-reverse programme we investigated the variations in the conception rates of sows over the corresponding one-year-period. In sows inseminated with semen from boars with natural photoperiod an annual pattern resulted which had the same trend as shown in fig. 11. The differences between highest and lowest conception rates however were only 3.6 %. Inseminations from light-programme boars changed the annual pattern of conception rates. Thus (compare to fig. 11) significant differences were found in March, May and August (see table 1).

TABLE 1   Non-return evaluation of 65 429 inseminations from natural-photo-period boars (N) and 63 782 inseminations from light-reverse-boars (L)

Month with significant differences (p= 0.05) between groups	degree of difference (%)	in favour of
March	1.05	L-boars
May	1.25	N-boars
August	1.20	L-boars

This indicates that a minor male-influence exists which is dependant on the photoperiod and is not explained by the number of sperms per insemination.

Seasonal pattern of the litter size

Many investigations deal with seasonal variations of the litter size (e. g. Legault et al., 1975 a, b; Lutter and Hühn, 1980; Welp, 1982; Coffey, et. al., 1983; Noguera et al., 1983). In most cases only a tendency for decreased litter sizes resulting from inseminations in summer was found. Other studies did not reveal such a seasonal variation in litter size (Love, 1978; Paterson et al., 1978; Keindorf and Plescher, 1981).

When evaluating a total of 9 081 litters born during one year we found a annual pattern which is shown in fig. 12. The difference in the average litter size resulting from inseminations in July (9.9 piglets) or November is 0.69 piglets. This difference is highly significant (p < 0.001). Similarly the litter size resulting from inseminations in November is significantly elevated compared to litter sizes from AI's in January until August. Even if the pattern in fig. 12 is surprisingly identical to others (e. g. fig. 1 or 11) the dependency on the photoperiod so far is not proven.

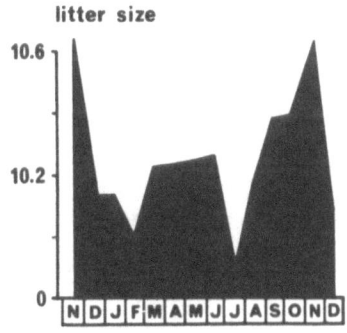

litter size

Fig. 12   Annual pattern of the litter size resulting from inseminations during the month indicated (Mezger, 1984).

REFERENCES

Aumaitre, A., Dagorn, J., Legault, C. and Le Denmat, M. 1976. Influence of farm management and breed type on sow's conception-weaning interval and productivity in France. Livestock Prod. Sci. 3, 75-83.

Bevier, G.W. and Backstrom, L. 1981. Seasonal infertility pattern during 1978 in 22 swine herds in Iowa and Nebraska, USA. Theriogenology 16, 321.

Cameron, R.D.A. 1980. The effect of heat stress on reproductive efficiency in breeding pigs. Veterinary Annual (UK) 20, 259-264.

Cameron, R.D.A. and Blackshaw, A.W. 1980. The effect of elevated ambient temperature on spermatogenesis in the boar. J. Reprod. Fert., 59, 173-179.

Claß, A. 1984. Untersuchungen über Testosteronkonzentrationen im Gesamtsamen von Ebern: Zeitspanne bis zur Reaktion auf geänderte Lichtverhältnisse. Diplomarbeit an der Universität Hohenheim, Fachgebiet Tierhaltung.

Claus, R., Schopper, D. and Wagner, H.-G. 1983. Seasonal effect on steroids in blood plasma and seminal plasma of boars. J. Steroid Biochem. 19, 725-729.

Claus, R., Schopper, D., Wagner, H.-G. and Weiler, U. 1984 a. Photoperiodic influences on reproduction of domestic boars. I. Light influences on testicular steroids in peripheral blood plasma and seminal plasma. Zbl. Vet. Med. A., in press.

Claus, R., Weiler, U. and Wagner, H.-G. 1984 b. Photoperiodic influences on reproduction of domestic boars. II. Light influences on semen characteristics and libido. Zbl. Vet. Med. A., in press.

Claus, R., Schelkle, G. and Weiler, U. 1984 c. Erste Versuche zur Verbesserung der Fruchtbarkeit von Sauen im Sommer durch ein Lichtprogramm. Zuchthyg. 19, 49-56.

Coffey, M.T., Olsen, T.A. and Combs, G.E. 1983. Effects of season and parity on sow productivity. Proc. 28th Anim. Swine Field Day, Florida, Oct. 3, 1983, 28-33.

Corteel, J.M., Signoret, J.P. and Du Mesnil du Buisson, F. 1964. Variations saisonniéres de la réproduction de la truie et facteurs favorisant l'anoestrus temporale. Proc. 5th Int. Congress Animal Production A.I., Trento, Italy, 536-540.

Van der Heyde, H., Lievens, R., Van Nieuwerburgh, G. and Doorme, H. 1974. Réproduction des truies en fonctions de diverses durées de lactation. I. Intervalle sevrage-oestrus. 2. Pourcentage de gestation. Revue de l'Agriculture, Brussels, 27, 1-37.

Herak, M. 1974. Rate of farrowing variation of sows covered during various months of the year. Proc. Int. Congress. I.P.V.S., Lyon, 2-3.

Hurtgen, J.P. and Leman, A.D. 1978. Seasonal breeding patterns in parous sows: a slaughterhouse survey. Vth Worldinternational pig veterinary society congr.

Hurtgen, J.P. and Leman, A.D. 1981 a. The seasonal breeding pattern of sows in seven confinement herds. Theriogenology, 16, 505- 511.

Hurtgen, J.P. and Leman, A.D. 1981 b. Effect of parity and season of farrowing on the subsequent farrowing interval of sows. The Veterinary Record, 108, 32-34.

Hurtgen, J.P., Leman, A.D. and Crabo, B., 1980. Seasonal influence on estrous activity in sows and gilts. J. Am. Vet. Med. Association 176, 119-123.

Jaussiaux, M. 1964. Contribution a l'étude des variations saisonniéres de la fécondité en insémination artificielle porcine. Vth Int. Cong. Anim. Reprod. Artif. Insem. Trento, 476-480.

Keindorf, A. and Plescher, W. 1981. Der Jahreszeiteinfluß auf die Fruchtbarkeitsleistungen der Schweine unter besonderer Berücksichtigung der Sommermonate. Monatshefte für Veterinärmedizin 36(9), 324-330.

Legault, C., Aumaitre, A. and Du Mesnil du Buisson, F. 1975 a. The improvement of sow productivity, a review of recent experiments in France. Livestock Prod. Sci. 2, 235-246.

Legault, C., Dagorn, J. and Tastu, D. 1975 b. Effets du mois de mise bas, du numéro de portée et du type génétique de la mére sur les composantes de la productivité de la trui dans les élevages francais. Journées Rech. Porcine en France, XLIII-LII.

Love, R.J. 1978. Definition of a seasonal infertility problem in pigs. Veterinary Record 103, 443-446.

Lutter, K. and Hühn, U. 1980. Untersuchungen über jahreszeitliche Schwankungen der Sauenfruchtbarkeit. Monatshefte für Veterinärmedizin 35, 819-822.

Martinat-Botte, F., Dando, P., Gautier, J. and Terqui, M. 1981. Variations saisonniéres de la taille de la portée en relation avec celles du taux d'ovulation et de la mortalité embryonnaire chez la truie. Journées Rech. Porcine en France, 269-276.

Mauget, R. 1982. Seasonality of reproduction in the wild boar. In: Control of pig Reproduction. Eds.: Cole, D.J.A., Foxcroft, C.R.; Butterworths, London, 509-526.

Mazzarri, G., du Mesnil du Buisson, F. and Ortavant, R. 1968. Action de la température et de la lumiére sur la spermatogenése, la production et le pouvoir fécondant du sperme chez le verrat. Proc. VIth Int. Cong. Anim. Reprod. Artif. Insem., Paris. Vol. I, 305-308.

Mezger, K. 1984. Saisonale Schwankungen des Fruchtbarkeitsgeschehens beim weiblichen Hausschwein - Untersuchung anhand von Daten einer Besamungsstation und eines Praxisbetriebes -. Diplomarbeit Universität Hohenheim, Fachgebiet Tierhaltung.

Moore, C.P., Dutt, R.H., Hays, V.W. and Cromwell, G.L. 1971. One-day vs. 14-day flushing and seasonal effects on reproduction in gilts. J. Anim. Sci. 33, 261.

Noguera, J.L., Felgines, C. and Legault, C. 1983. Evolution de 1972 á 1981 des composantes de la productivité numérique des truies dans 325 troupeaux francais. Journée Rech. Porcine en France, 37-52.

Ortavant, W., Mauleon, P. and Thibault, C. 1964. Photoperiodic control of gonadal and hypophyseal activity in domestic animals. Annals New York Academy of Sciences, 117, 157-193.

Paterson, A.M., Barker, I. and Lindsay, D.R. 1978. Summer infertility in pigs: its incidence and characteristics in an Australian commercial piggery. Austr. J. Exper. Agr. Anim. Husbandry 18, 698-701.

Peter, W., Franck, Ch., Mudra, K. and Ueckert, H. 1981. Der Einfluß der Aufzucht von Jungebern in zentralen Stationen auf Besamungseignung und Spermaproduktionsvermögen. Tierzucht, 35, 92-95.

Roller, W.L. and Stombaugh, D.P. 1974. The influence of environmental factors on reproduction of livestock. Proc., Int. Livestock Environment Symposium, Lincoln, 31-50.

Schindler, A. 1980. Saisonalität des Fortpflanzungsgeschehens beim Eber? Untersuchungen anhand der Praxisergebnisse einer Besamungsstation. Diplomarbeit TU München, Weihenstephan.

Signoret, J.-P. 1967. Durée du cycle oestrien et de l'oestrus chez la truie, action du benzoate d'oestradiol chez la femelle ovariectomisée. Annales Biol. Anim. Biochm. Biophys., Paris, 7, 407-421.

Steinbach, J. 1976. Reproductive performance of high-producing pigs under tropical conditions. World Anim. Review 19, 43-47.

Stork, M.G. 1979. Seasonal reproductive inefficiency in large pig breeding units in Britain. Veterinary Record 104, 49-52.

Swierstra, E.E. 1968. Cytology and duration of the cycle of the seminiferous epithelium of the boar; duration of spermatozoan transit through the epididymis. Anatomical Rec., 161, 171-185.

Tomes, G.T. and Nielson, H.E. 1979. Seasonal variations in the reproductive performance of sows under different climatic conditions. World Review of Anim. Prod. 15(1), 9-19.

Welp, C. 1982. Die Anwendung biotechnischer Maßnahmen zur Steigerung der Fruchtbarkeitsleistung von Sauen unter Feldbedingungen. Dissertation, Georg-August Universität Göttingen.

Wettemann, R.P. and Desjardins, C. 1979. Testicular function in boars exposed to elevated ambient temperature. Biol. Reprod., 20, 235-241.

Williamson, P., Hennessy, D.P. and Cutler, R. 1980. The use of progesteron and oestrogen concentrations in the diagnosis of pregnancy and in the study of seasonal infertility in sows. Australian J. Agric. Research 31(1), 233-238.

Wrathall, A.E. 1977. Reproductive failure in the pig: diagnosis and control. Veterinary Record 100, 230-237.

DISCUSSION

Chairman: Roche, J.F. (Ireland)

The first two papers dealt with seasonal anestrus in the mare. B. Bour (France) indicated that the mare is a seasonal breeder, responding to long days. Experiments showed that 16 h light advances the breeding season. Attempts to define a photosensitive phase indicated that a 1 h light pulse given 9.5 to 10.5 h after dusk was optimal in advancing the breeding season. The constraints on photoperiod were summarized as light being present 9.5 to 10.5 h after dark, the total length of the light period was not important and a minimal period of darkness was required. Work on the hormonal induction of ovulation indicating either HCG or equine LH and FSH were effective in the late transitional phase of anestrus, but there was an interaction between body condition and hormone stimulation.

W.R. Allen (U.K.) reviewed the different types of anestrus in the mare. Seasonal anestrus varies with breed and nutritional level but a well defined anestrous period from November to February exists. In the transitional phase from anestrus to cyclicity, some mares entered a prolonged estrus lasting in some cases for more than 30 days. This prolonged estrus can be terminated by synthetic progesterone treatment for 8 - 10 days. Lactational anestrus occurs to a variable extent, particularly in mares foaling early in the season, and this can be reduced by exposing these mares to 16 h light in late pregnancy. Prolonged diestrus occurs during the breeding season which appears to be due to a failure of complete regression of the corpus luteum. This condition can be inferred where non-pregnant mares fail to return to estrus and can be confirmed by progesterone measurement. Pregnancy anestrus can occur if pregnant mares lose the embryo between days 40 and 120. Because of formation of the endometrial cups and continued secretion of PMSG, these non-pregnant mares fail to return to estrus for long periods of time.

Two papers on seasonal reproduction in the pig were also presented. R. Mauget (France) discussed the clearly defined anestrus period in summer in wild boars and sows. Nutrition can advance th onset of the breeding season by 2 - 3 weeks, and temperature has very little role to play. Matings take place in the autumn/winter months with a single annual farrowing occurring in spring.

R. Claus (F.R.G.) also documented seasonal changes in reproduction in domesticated boars and sows but, in contrast to the wild pig, a true anestrous period is absent. In the boar the number of motile sperm per ejaculate, testosterone concentrations in seminal plasma and libido were higher in winter. The application of artificial light programs (short days in summer) can increase sexual function in the boar during the summer. In the sow, the interval from weaning to first estrus, conception rates and litter sizes are influenced by season. By decreasing daily light exposure by 20 min per week, starting in May, a significant reduction in weaning to estrus interval occurred, and litter size was also increased. Therefore, adverse reproduction efficiency in sows and boars may be overcome to some extent by appropriate changes in photoperiod, and this is an area of continued development in future. However, genetic variation and nutritional effects are important factors to be considered in discussing seasonal effects on reproduction.

SESSION III

LACTATIONAL ANESTRUS IN THE PIG

Chairman: R. Mauget

HYPOTHALAMIC-PITUITARY-OVARIAN RESPONSES TO WEANING - RETROSPECTIVE EVIDENCE FOR THE CAUSE OF LACTATIONAL ANOESTRUS?

George R. Foxcroft

AFRC Research Group on Hormones and Farm Animal Reproduction,
University of Nottingham Faculty of Agricultural Science,
Sutton Bonington, Loughborough, LE12 5RD, U.K.

ABSTRACT

Evidence from the study of endogenous hormone changes at the time of weaning in the sow suggest that even during lactation, active secretion of both LH and FSH occurs. The extent to which this escape from the inhibitory effects of suckling and lactation occurs varies between sows and is related to both the response of the endocrine system to weaning and to the weaning to oestrus interval. The length of lactation also affects the endocrine status of the sow. A gradual recovery of tonic gonadotrophin secretion and in the responsiveness of the positive feedback system occurs, and the magnitude of the preovulatory LH surge after weaning increases, with time of weaning post-partum. Sows returning to oestrus immediately after weaning have a characteristic increase in basal LH secretion; in sows returning to oestrus later, a significant increase in episodic LH secretion occurs but not a sustained rise in baseline. Plasma FSH levels are related to the development of steroidogenic activity and to ovulation rate after weaning. The mechanisms controlling the inhibition of reproduction during lactation are uncertain, but a primary block at the hypothalamic level is evident. Both prolactin and oxytocin may mediate such effects and steroids may also exert important modulatory effects as lactation progresses.

INTRODUCTION

An investigation of the endocrine responses to weaning, when the inhibitory effects of suckling and lactation are removed, is relevant to a discussion of lactational anoestrus for two reasons. Firstly, it is logical to assume that the changes in hypothalamic-pituitary-ovarian activity seen after weaning will indicate the locus of action of the direct and indirect inhibitory effects of lactation on reproduction. Study of the weaned sow therefore offers a model in which to study the endocrinology of the sow at a time when follicular growth and steroidogenesis is stimulated, as evidenced by the early return to oestrus in the majority of sows in conventional management systems. During this period of ovarian stimulation each component of the endocrine response to weaning should be open to manipulation experimentally using a range of physiological and pharmacological techniques. These may inhibit or enhance the synthesis and/or release of specific hormones or alternatively involve agonists, antagonists or antibodies to block the binding or intracellular activity of hormones at their target tissues.

Secondly, variability in the time to return to oestrus in the weaned sow, even in highly fertile herds, is normal. This in turn presumably reflects variability in the function of one or more component(s) of the endocrine axis, viz. differences in the stimulatory activity of hormones or in the demise of inhibitory mechanisms after weaning, or differential sensitivity of target tissues to stimulation. Experimental evidence for the endocrine basis of variable ovarian stimulation after weaning should also therefore provide some indication of the mechanisms that inhibit follicular development during lactation, on the assumption that at least some of the mechanisms controlling lactational and post-weaning anoestrus are common. In this paper therefore data will be presented which describes endogenous endocrine changes in the weaned sow and permits some conclusions to be made as to the possible cause of variability in ovarian activity in such animals.

ENDOGENOUS ENDOCRINE CHANGES IN WEANED SOWS
Gonadotrophin releasing hormone (GnRH)

Although no evidence is available on the actual secretion of releasing hormones in the pig, Cox and Britt (1981) have reported that the hypothalamic content of GnRH shows a significant increase within 60 hours of weaning. Although the relationship between this increase and the actual release of GnRH and subsequently LH and FSH is unclear, these data at least indicate changes at the level of the hypothalamus as a consequence of weaning. On the assumption that pulses of GnRH from the hypothalamus are temporarily linked to episodes of LH release, the variability in episodic LH secretion during lactation and after weaning (see next section) suggest that similar changes in hypothalamic activity occur and that this is one locus of action of the inhibitory effects of suckling and lactation.

Gonadotrophins
Luteinizing hormone (LH): The earliest direct evidence for changes in gonadotrophin secretion in the lactating/weaned sow was derived from bioassay of pituitary LH and FSH content (Lauderdale et al., 1965; Melampy et al., 1966; Crighton and Lamming, 1969). Low or declining pituitary LH levels during lactation followed by a significant increase after weaning

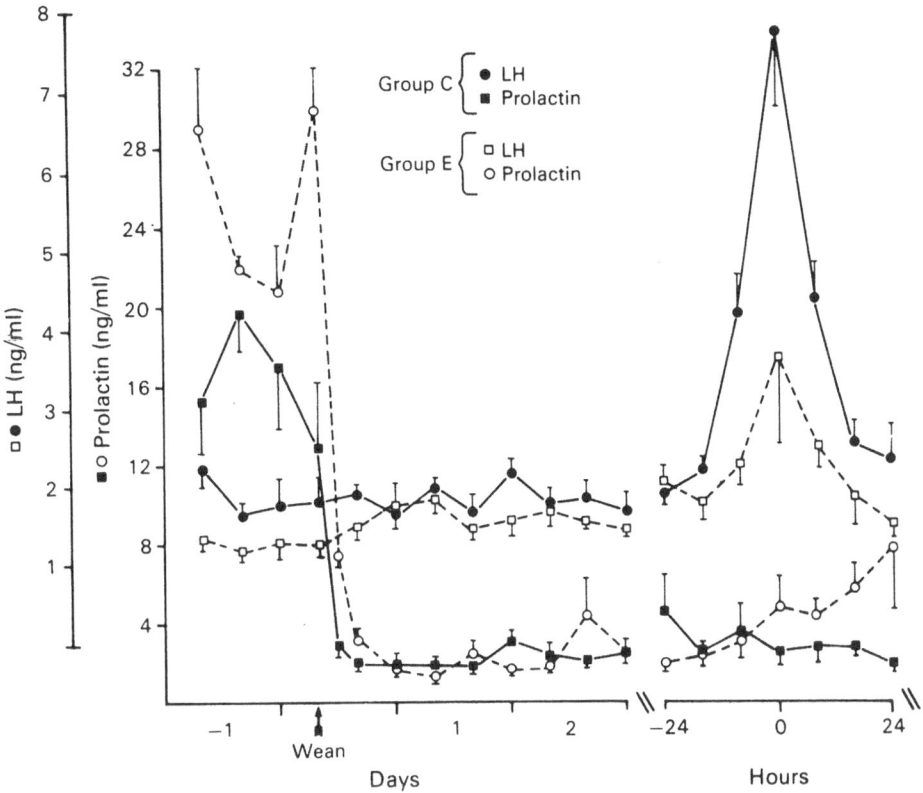

Fig. 1. Mean + s.e.m. plasma LH and prolactin concentrations in sows from -1 to 2 days after weaning and at the time of the preovulatory LH surge. Group E weaned at 10 days and Group C at 35 days of lactation. (From Kirkwood et al., 1984).

was interpreted as evidence for the suppression of both LH synthesis and release during lactation. An increase in pituitary LH after weaning has subsequently been confirmed by Cox and Britt (1981) using radioimmunoassay (RIA) techniques. The lack of an increase in both pituitary LH content (Crighton and Lamming, 1969) and plasma LH (Parvizi et al., 1976; Stevenson et al., 1981) following ovariectomy also suggests that the suppression of LH release during lactation is not ovarian dependent, but due to a more direct effect of suckling and lactation at the hypothalamic-pituitary level.

136

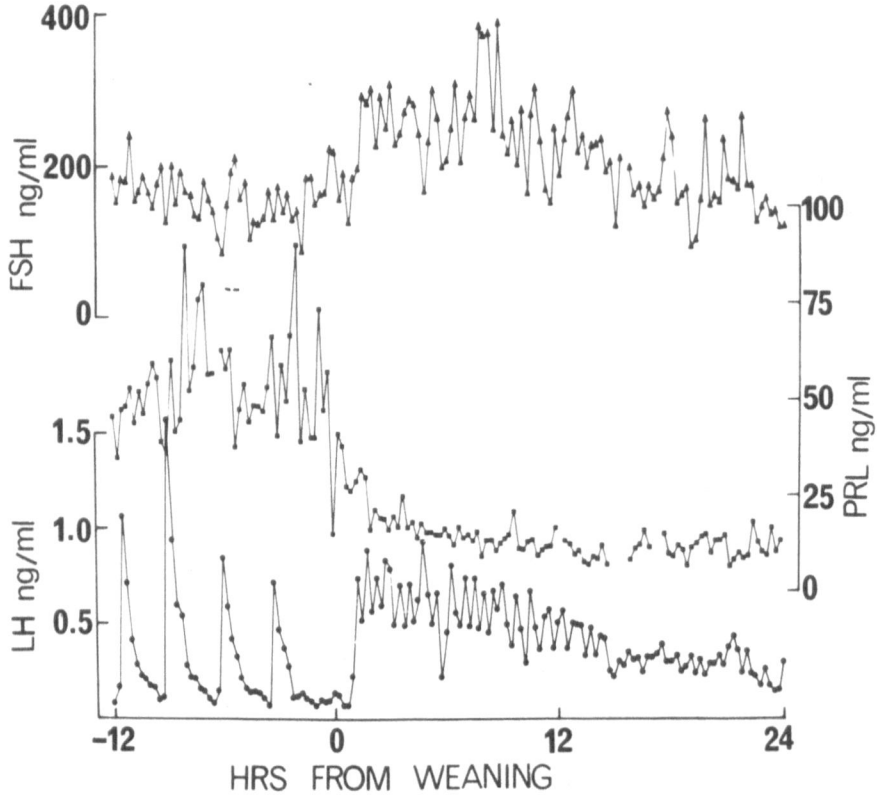

Fig. 2. Plasma LH, FSH and prolactin measured continuously at 15 min intervals in a sow before and after weaning at 21 days of lactation. (From Shaw, 1984).

These data have been considerably expanded using validated RIA techniques to measure plasma LH directly and apparent differences in reported levels of LH are probably mainly related to the sampling frequency used by different authors. Data from experiments using relatively infrequent sampling suggests that LH secretion was either absent, or at very low levels, during lactation (Aherne et al., 1976; Parvizi et al., 1976; Booman and Van de Wiel, 1980), nevertheless an increase in LH levels as lactation progressed has also been reported (Stevenson and Britt, 1980; Stevenson et al., 1981; Kirkwood et al., 198? see Fig. 1). With frequent sampling during lactation it has been conclusively demonstrated that episodic release of LH is present from at least 21 days post-partum (Edwards and Foxcroft, 1983; Shaw, 1984). In

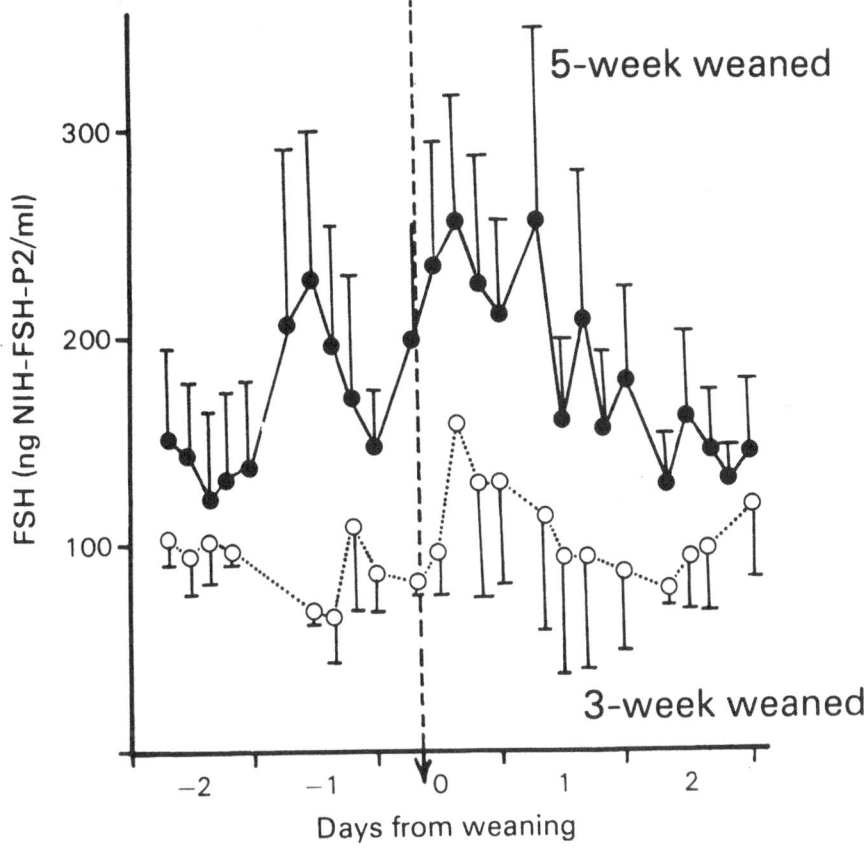

Fig. 3. Mean plasma FSH measured in groups of sows before and after weaning at 3 or 5 weeks of lactation, using a 4-hourly sampling frequency. (From Edwards and Foxcroft, 1983).

the latter study a significant inverse relationship was also established between the mean LH levels before weaning and the weaning to oestrus interval, suggesting that the level of episodic LH secretion during lactation has a critical influence on the ovarian response to weaning.

There are consistent reports of an eventual increase in plasma LH after weaning, although the time at which this increase became significant was again affected by sampling frequency (Aherne et al., 1976; Parvizi et al., 1976; Van de Wiel et al., 1979; Stevenson and Britt, 1980; Cox and Britt, 1981; Edwards and Foxcroft, 1983; Kirkwood et al., 1984). Using continuous frequent sampling at the time of weaning Shaw (1984) has demonstrated that a consistent and immediate increase in episodic LH

secretion occurs in sows weaned at 21 days of lactation, irrespective of
the subsequent weaning to oestrus interval. The pattern of LH secretion
was however characteristic of the time to oestrus. In sows with an early
return to oestrus (<5 days) the pattern of LH secretion was such as to
produce an immediate and continuous elevation of basal LH secretion, as
shown in Fig. 2. This pattern of LH secretion is suggestive of a high
frequency pulsatile release of GnRH; also in comparison to the
characteristics of episodic LH release and to varying responsiveness to
exogenous GnRH stimulation during the oestrous cycle, these data also
indicate possible oestrogen-dependent modulation of LH secretion. It
remains to be determined however whether this particular pattern of LH
release after weaning is the stimulus for, or a consequence of, ovarian
steroidogenic activity. Certainly the timing of the preovulatory LH surge
in early returning sows indicates that peripheral levels of oestradiol are
able to produce an effective positive feedback stimulus within 24 hours of
weaning. In sows returning to oestrus between five and nine days after
weaning the increase in mean LH levels was associated with a sustained
increase in both episodic amplitude and frequency but not with any
consistent increase in LH baseline. Althouugh an initial increase in
episodic LH secretion occurred in sows with a weaning to oestrus interval
of ten days or more, this was not sustained.

The characteristics of the pre-ovulatory LH surge after weaning do
not appear to be affected by the weaning to surge interval in sows weaned
at a similar stage of lactation; however an increase in lactation length
is associated with an increase in the magnitude of this preovulatory LH
surge as shown by Edwards and Foxcroft (1983) and Kirkwood et al. (1984).
These data confirm evidence for a gradual recovery in the responsiveness
of the oestrogen positive feedback mechanism during lactation (Elsaesser
and Parvizi, 1980). As discussed later, the variable responsiveness of
the positive feedback mechanism is not apparently related to differences
in the pre-surge increase in oestradiol secretion; however changes in
steroid feedback activity during lactation cannot be excluded as being
functionally related to the recovery of the positive feedback system.
Preliminary evidence also suggests that a recovery of pituitary
responsiveness to GnRH stimulation is not an inherent component of the
recovery of the positive feedback mechanism, as similar LH responses to
GnRH treatment have been reported at different stages of lactation (Van de
Wiel et al., 1978; Stevenson et al., 1981).

Generally therefore a gradual recovery of tonic LH secretion and in the responsiveness of the LH surge mechanism occurs during lactation. The extent of this recovery in LH secretion is related to both the characteristics of episodic LH release and to the rate of ovarian follicular development -after weaning.

Follicle stimulating hormone (FSH): High pituitary FSH potencies have been reported during lactation using both bioassay (Lauderdale et al., 1965; Crighton and Lamming, 1969) and RIA (Cox and Britt, 1981). In conjunction with morphological evidence indicating an overall suppression of follicular growth during lactation, these data were interpreted as evidence for active synthesis, but not release, of FSH prior to weaning. The direct determination of plasma FSH levels suggest however that active secretion of FSH occurs prior to weaning and increases as lactation progresses (Aherne et al., 1976; Stevenson et al., 1981; Duggan et al., 1982; Edwards and Foxcroft, 1983; see Fig. 3). Several of these authors also emphasize that FSH levels in late lactation are similar to those seen in the late luteal phase of the oestrous cycle and a lack of FSH secretion may not be a primary cause of arrested follicular development during lactation. Nevertheless plasma FSH levels increased significantly at weaning in a number of studies (Aherne et al., 1976; Cox and Britt, 1981), although Edwards and Foxcroft (1983) observed no increase in FSH secretion in several sows that returned to oestrus normally. This variability in the FSH response to weaning was confirmed in the studies of Shaw (1984) who also however established that when manipulation of suckling intensity was used to enhance follicular development during lactation, plasma levels of FSH were significantly and positively correlated to the level of granulosa cell aromatase activity. Variations in the secretion of FSH may still therefore be important in the control of follicular development during lactation.

The magnitude of the rise in plasma FSH at the time of the preovulatory LH surge was significantly affected by the time of weaning (Edwards and Foxcroft, 1983), however as the lack of such a rise in FSH in early weaned sows was not associated with a failure in ovulation it appears that a pre-ovulatory rise in FSH is not essential to the ovulatory process. However in an analysis of the relationships between endogenous hormone changes and reproductive function Shaw (1984) established a positive relationship between mean plasma FSH levels and ovulation rate.

Perhaps predictably therefore, direct evidence suggests that suckling and lactation suppress gonadotrophin secretion. However, as lactation progresses this suppression is far from complete and active synthesis and release of LH and FSH are present. This gradual development of gonadotrophin activity is paralleled by a gradual increase in follicular development and by a decrease in the proportion of atretic follicles within the ovary (Palmer et al., 1965a, b; Lauderdale et al., 1965; Crighton, 1966; Kunavongkrit et al., 1982). Evidence is still lacking however for a functional relationship between differences in gonadotrophin secretion during lactation and the development of ovarian activity.

Oxytocin: Although the release of oxytocin has long been associated with the milk ejection in mammals, direct determinations of plasma oxytocin in lactating sows have only recently been reported (see Ellendorff, 1984). Evidence for the possible role of oxytocin as a mediator of the inhibitory effects of suckling on ovarian function is also limited. Peters et al. (1969) demonstrated that mammillectomy in sows was associated with enhanced follicular development even if piglets were present; furthermore although oxytocin injections (every 2h for 7 days from parturition) in these mammillectomized sows resulted in high pituitary FSH potencies compared to untreated controls, litter presence in addition to oxytocin treatment was required to inhibit follicular development. Further evidence to implicate oxytocin in the suppression of follicular development will be presented in the next paper, but the locus of action of this neuro-peptide within the hypothalamo-pituitary-ovarian axis is uncertain.

Prolactin: Discussion on this topic will be limited, as two papers in this session will describe the role of prolactin in the lactating-weaned sow in some detail. Considering only the post-weaning period, a rapid and predictable decline in plasma prolactin has been observed consistently after weaning (see Figs. 1 and 2) and there is no evidence to date that hyperprolactinaemia is functionally related to cases of post-weaning anoestrus (van Landeghem and Van de Wiel, 1978; Bevers et al., 1978; Booman and Van de Wiel, 1980; Edwards and Foxcroft, 1983; Shaw, 1984).

As prolactin secretion is high during lactation it is however entirely possible that hypothalamic releasing or inhibiting factors, or prolactin itself, may exert an inhibitory effect on ovarian functiion and that this inhibition may be effected at any level of the hypothalamic-pituitary-ovarian axis. It is important therefore to determine the relative importance of prolactin in the suppression of ovarian function during lactation and the locus of action of this hormone or its regulatory factors.

Oestradiol: The determination of the circulating levels of oestradiol-17β is of particular significance, as increased oestradiol secretion provides evidence for a resumption of ovarian steroidogenesis. A sustained rise in oestradiol is also a prerequisite for oestrus and ovulation and the onset of oestrus after weaning has been consistently associated with a preceeding surge in oestrogen in peripheral plasma (Ash and Heap, 1975; Aherne et al., 1976; Stevenson and Britt, 1980; Edwards and Foxcroft, 1983; Kirkwood et al., 1984).

In a number of studies however, the circulating levels of oestradiol in lactation were very variable (Stevenson et al., 1981; Edwards and Foxcroft, 1983; Kirkwood et al., 1984) and strongly suggest that even in the presence of suckling and lactation, active steroidogenesis is present in at least some sows. The mechanisms by which this occurs have yet to be elucidated, but even when oestradiol levels were elevated there was no evidence of subsequent oestrus or ovulation, suggesting that a dysfunction of the oestrogen feedback occurs centrally and still prevents the resumption of ovarian cycles.

DISCUSSION AND CONCLUSIONS

Based on evidence from the study of endogenous endocrine changes at weaning, the primary cause of lactational anoestrus appears to be a suppression of gonadotrophin secretion; however inhibition at the ovarian level may also exist. Even in the event of normal steroidogenic responses to a rise in endogenous gonadotrophins, oestrus and ovulation may still not occur in the lactating sow due to dysfunction of the oestrogen feedback mechanisms within the hypothalamic-pituitary unit.

One of the greatest problems to applying exogenous treatment to control reproduction is the considerable variability between animals in the degree of follicular development at any stage of lactation. This situation is projected to the weaned sow and evidence for the variability in follicular steroidogenesis in two sows 48h after weaning is shown in Fig. 4. If the physiological mechanisms controlling such differences were

142

understood, then considerable progress could be made to improve the reproductive efficiency of the sow.

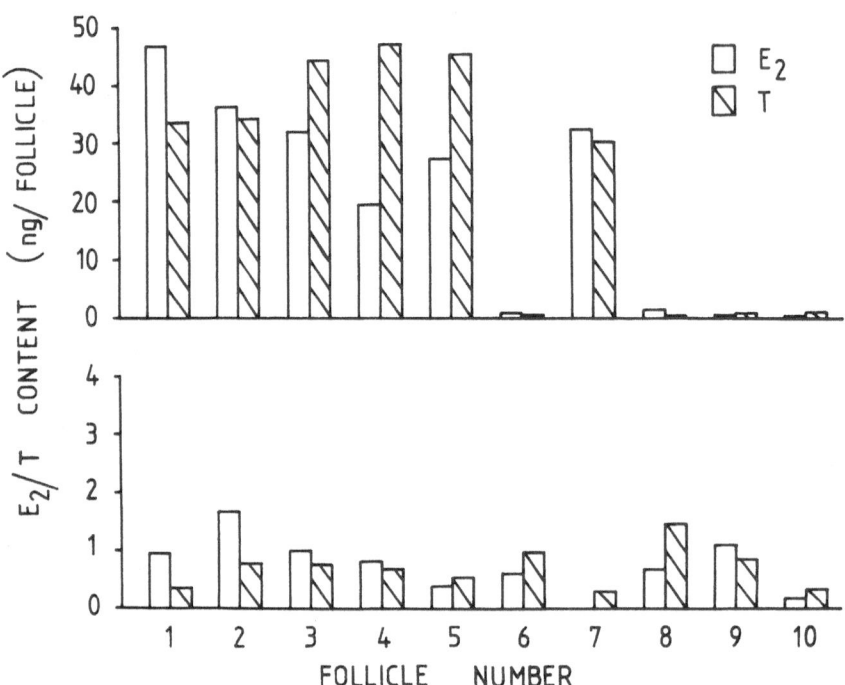

Fig. 4. Follicular fluid content of oestradiol-17β (E_2) and testosterone (T) in the largest ten follicles dissected from the ovaries of two sows weaned at approximately 3 weeks post-partum. (From Shaw, 1984).

REFERENCES

Aherne, F.X., Christopherson, R.J., Thompson, J.R. and Hardin, R.T. 1979. Factors affecting the onset of puberty, post-weaning estrus and blood hormone levels of Lacombe gilts. Can. J. Anim. Sci., 56, 681-692.

Ash, R.W. and Heap, R.B. 1975. Oestrogen, progesterone and cortico-steroid concentrations in peripheral plasma of sows during pregnancy, parturition, lactation and after weaning. J. Endocr., 64, 141-154.

Bevers, M.M., Willemse, A.H. and Kruip, Th.A.M. 1978. Plasma prolactin levels in the sow during lactation and the post-weaning period as measured by radioimmunoassay. Biol. Reprod., 19, 628-634.

Booman, P. and Van de Wiel, D.F.M. 1980. Lactational anoestrus in the pig: possible relationship with hyperprolactinaemia. Report B-157. Instituut voor Veeteeltkundig Onderzoek 'Schoonoord', Driebergseweg 10, Zeist, Holland.

Cox, N.M. and Britt, J.H. 1981. Relationship between endogenous GnRH and post-weaning endocrine events in sows. (Abstr.) Annual meeting of Southern Section, American Society of Animal Science, Atlanta, Georgia.

Crighton, D.B. 1966. The effects of lactation on reproduction in the pig. Ph.D. Thesis, University of Nottingham.

Crighton, D.B. and Lamming, G.E. 1969. The lactational anoestrus of the sow, the status of the anterior pituitary-ovarian system during lactation and after weaning. J. Endocr. 43, 507-519.

Duggan, R.T., Bryant, M.J. and Cunningham, F.J. 1982. Gonadotrophin, total oestrogen and progesterone concentrations in the plasma of lactating sows with particular reference to lactational anoestrus. J. Reprod. Fert., 64, 303-313.

Edwards, S. and Foxcroft, G.R. 1983. Endocrine changes in sows weaned at two stages of lactation. J. Reprod. Fert. 67, 161-172.

Ellendorff, F. 1984. The neurohypophyseal complex. Proceedings of 10th International Congress on Animal Reproduction and Artificial Insemination, Vol. IV, pp. I-18-I-28.

Elsaesser, F. and Parvizi, N. 1980. Partial recovery of the stimulatory oestrogen feedback action on LH release during late lactation in the pig. J. Reprod. Fert., 59, 63-67.

Foxcroft, G.R. and Van de Wiel, D.F.M. 1982. Endocrine control of the oestrous cycle. In 'Control of Pig Reproduction', (Eds. D.J.A. Cole and G.R. Foxcroft). (Butterworth, London). pp. 161-177.

Kirkwood, R.N., Lapwood, K.R., Smith, W.C. and Anderson, I.L. 1984. Plasma concentrations of LH, prolactin, oestradiol-17β and progesterone in sows weaned after lactation for 10 or 35 days. J. Reprod. Fert., 70, 95-102.

Kunavongkrit, A., Einarsson, S. and Settergren, I. 1982. Follicular development in primiparous lactating sows. Anim. Reprod. Sci., 5, 47-56.

Lauderdale, J.M., Kirkpatrick, R.L., First, N.L., Hauser, E.R. and Casida, L.E. 1965. Ovarian and pituitary gland changes in periparturient sows. J. Anim. Sci., 24, 1100-1103.

Melampy, R.M., Hendricks, D.M., Anderson, L.L., Chen, C.L. and Schultz, J.R. 1966. Pituitary LH and FSH concentrations in pregnant and lactating pigs. Endocrinology, 78, 801-804.

Palmer, W.M., Teague, H.S. and Venzke, W.G. 1965. Macroscopic observations on the reproductive tract of the sow during lactation and early post-weaning. J. Anim. Sci., 24, 541-545.

Parvizi, N., Elsaesser, F., Smidt, D. and Ellendorff, F. 1976. Plasma LH and progesterone in the adult female pig during the oestrous cycle, late pregnancy and lactation, and after ovariectomy and pentobarbitone treatment. J. Endocr., 69, 193-203.

Peters, J.B., First, N.L. and Casida, L.E. 1969. Effects of pig removal and oxytocin injections on ovarian and pituitary changes in mammillectomized post-partum sows. J. Anim. Sci., 28, 537-541.

Shaw, H.J. 1984. Control of ovarian function in lactating and weaned sows. Ph.D. Thesis, University of Nottingham.

Stevenson, J.S. and Britt, J.H. 1980. Luteinizing hormone, total

144

    oestrogen and progesterone secretion during lactation and after
    weaning in sows. Theriogenology, 14, 453-463.
Stevenson, J.S., Cox, N.M. and Britt, J.H. 1981. Role of the ovary
    in controlling luteinizing hormone, follicle stimulating
    hormone and prolactin secretion during and after lactation in
    pigs. Biol. Reprod. 24, 341-353.
Van de Wiel, D.F.M., Van Landeghem, A.A.J., Bevers, M.M. and
    Willemse, A.H. 1978. Endocrine control of pituitary and
    ovarian function during lactation in the sow. In 'Comparative
    Endocrinology' (Eds. P.J. Gaillard and H.H. Boer). (Elsevier/
    North Holland Press). p.378.
Van de Wiel, D.F.M., Van Landeghem, A.A.J., Willemse, A.H. and
    Bevers, M.M. 1979. Endocrine control of ovarian function after
    weaning in the domestic sow. J. Endocr., 80, 69P.
Van Landeghem, A.A.J. and Van de Wiel, D.F.M. 1978. Radioimmuno-
    assay for porcine prolactin: plasma levels during lactation,
    suckling and weaning and after TRF administration. Acta
    Endocrinol., 88, 653-667.

ENDOCRINE INTERACTIONS DURING LACTATIONAL ANESTRUS IN SOWS

F. Ellendorff, F. Elsaesser, N. Parvizi, D. Smidt

Institut für Tierzucht und Tierverhalten (FAL), Mariensee
3057 Neustadt 1, Federal Republic of Germany

ABSTRACT

The contribution deals with two aspects of lactational anestrus in the sow: (1) the depth of anestrus weakens with length of lactation, (2) the suckling stimulus and oxytocin release during lactational anestrus.

There is firm evidence that prolongation of lactation makes the anestrous sow increasingly susceptible to manipulations that aim at termination of lactational anestrus. These include removing of the suckling stimulus by weaning or treatment with PMSG/HCG. Circulating LH and prolactin levels reflect the changing status.

As to the suckling stimulus, a number of components have now been defined which suggests that one or more may be responsible for lactational anestrus. The relatively high initial suckling interval becomes prolonged as lactation advances, so does the secretory frequency of oxytocin and the total amount of oxytocin released. It needs to be determined whether escape from inhibition of the sows reproductive activity occurs at the level of the CNS or is due to some direct action of oxytocin at the ovarian level.

INTRODUCTION

Lactational anestrus in the pig is an acyclic post partum stage extending beyond the puerperal period and is maintained by the presence of several suckling piglets. The puerperal period usually lasts for about three weeks after parturition. Removal of piglets results in resumption of cyclicity about 5 - 7 days after weaning. Even though this sequence seems rather ridgid at first sight, part one of the following discussion will show, it is not, but is related to a sequential escape from inhibitory influences on the CNS-pituitary-gonadal system. The second part will describe suckling-related endocrine phenomena that are associated with the sequential disinhibition and may serve as a possible cause for lactational anestrus in sows.

Depth of anestrus weakens with length of lactation

If, contrary to accepted husbandry practice, piglets are not weaned and lactating sows are kept with a boar, sows

146

eventually become pregnant, even though piglets may still be
suckling. These observations clearly indicate, that lactation and nursing do not inhibit resumption of cyclic activity and ovulation indefinitely. More than one piglet is
necessary to maintain lactation and to prevent resumption of
cyclic activity (Parvizi et al., 1976).

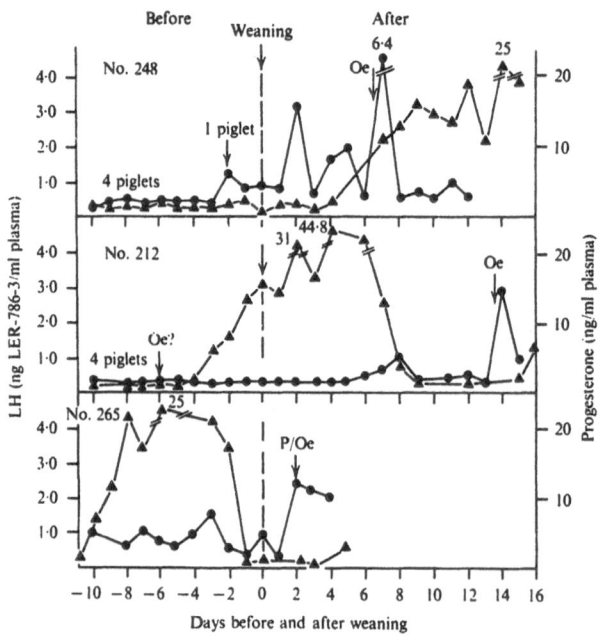

Fig. 1 Plasma LH (•) and progesterone (▲) prior and after weaning in
three pigs. Piglets were weaned at 14, 28 and 24 days (from top).
Sow 212: Piglets removed for 24 h on day 17 of lactation, Sow 265 had only
2 piglets left. (From Parvizi et al., 1976).

Thus it may be assumed, that inhibition of the CNS-pituitary-
ovarian-axis weakens as lactation progresses. When ovarian
activity is monitored by circulating levels of gonadal steroids during the first 4 - 6 weeks of lactation, both plasma
progesterone and plasma estrogens are close to the lower limits of detection (Ash & Heap, 1975; Parvizi et al., 1976;
Stephenson, Cox, Britt, 1981). Plasma LH remains at low levels

as well, but changes in later lactation (Parvizi et al., 1976). Prolactin is secreted at a relatively high rate throughout lactation but declines at later stages which has led to the assumption that high prolactin levels are the cause of lactational anestrus (van Landeghem, van de Wiel, 1978; Bevers, Willemse & Kruip, 1978).

If piglets are weaned at 3 weeks of age, interval to final estrus amounts to almost 13 days, if nursing was interrupted for alternating 12 h periods for 48 h prior to weaning, then the interval to first postweaning estrus was only about 8 days (Britt & Levis, 1982).

There is considerable evidence that shorter suckling periods prolong the weaning to estrus interval and increase the variability, while longer lactation periods shorten the interval and decrease variability (Smidt et al., 1965; Dyck et al., 1979; Britt & Levis, 1982).

Since inhibition of cyclic activity is expressed at the ovarian and at the pituitary level, the responsiveness of both elements to various exogenous stimuli has been tested.

At the ovarian level, Pregnant Mare Serum Gonadotropin (PMSG) and Human Chorionic Gonadotropin (HCG) treatment indicates a gradual increase in responses as lactation progresses. Even though estrus and pregnancies may be induced by PMSG as early as 15 days post partum, significant differences in pregnancy rates are obvious when PMSG/HCG-treated sows of 14 - 24 days post partum are compared to sows of 24 - 45 days (Hodson et al., 1981). Not only length of lactation affects the response to PMSG, so does the number of piglets nursing. If, e.g. on day 10, litter size had been adjusted to 10 - 15 piglets and PMSG was injected, only 30% of sows displayed estrus, none was pregnant upon insemination. If, on the other hand, only 1 - 5 litters were left with the sow, then about 75% came into estrus and about 65% became pregnant upon insemination (Martinat-Botte, 1975). Thus, sensitivity of the ovary depends at least on duration of lactation and/or number of piglets suckling. The longer the lactation lasts and the fewer the number of piglets suckling, the more receptive is the ovary towards PMSG.

At the pituitary, the major trigger for increased FSH/LH is hypothalamic GnRH (LH/FSHRH). Discharge of GnRH and/or effectiveness of GnRH at the pituitary may be modified by circulating steroids with an essential estrogen surge prior to a preovulatory discharge of LH. GnRH-treatment of sows in week 1, 2 and 3 of lactation indicates on an individual base, that each of three groups displayed an increase in the area under LH-peak of more than two to more than four-fold when week one of lactation was compared to week three. Application of GnRH in week two took an intermediate position (Bevers et al., 1981). Other observations point to the same pattern (Cox & Britt, 1982), though less conclusive with respect to an increase in responsiveness of the pituitary with advancing length of lactation. Apparently the height of prolactin levels does not seem to influence the response (Bevers et al., 1981). One factor contributing to the changing responsiveness of the pituitary to GnRH could be a lack of estrogen sensitivity of the pituitary,even though only minute levels of estrogen are circulating during lactation. Treatment of 5 day lactating sows with estradiol dose levels, which elicit a positive surge of LH in peripubertal sows does not change plasma LH levels when compared to controls. The same procedure in 35 day old sows does induce a subsequent significant elevation of plasma LH levels followed by luteal activity in about 20% of the animals (Elsaesser & Parvizi, 1980).

From the foregoing it has become clear, that the depth of anestrus weakens as lactation length increases. Both, ovary and pituitary are sites, where depth of anestrus is expressed. Despite knowledge of the results of lactation and of reproduction associated with lactation, little is known on causative mechanisms.

TABLE 1. Effect of estradiol benzoate treatment on plasma LH levels
and luteal activity in lactating sows

Group	Means±SEM max.LH levels (ng/ml) 48-72 h after treatment	Percentage of animals with luteal activity+			
		After treatment		After weaning on day 49	
		8 days	13 days	10 days	17 days
Day 5 EB[1]	0.73 ± 0.13	0.0	0.0	62.5	83.3
Day 5 S[2]	0.83 ± 0.16	0.0	0.0	80.0	100.0
Day 35 EB	1.75 ± 0.27*	22.2	22.2	14.3	33.3
Day 35 S	0.95 ± 0.19	0.0	0.0	80.0	100.0

*P < 0.05 compared with Day 35 S 1) 60 µg/kg BW i.m.
+ Plasma progesterone levels > 5 ng/ml 2) Sesame oil
(From Elsaesser & Parvizi, 1980)

The suckling stimulus and oxytocin release during lactatio-nal anestrus

Striking parallelism exists between suckling-related
events and the depth of anestrus. It was already mentioned,
that the depth of anestrus is inversely related to the number
of piglets suckling, with a minimum of 2 or more piglets necessary to maintain lactational anestrus. Other events relate
to suckling frequency (nursing intervals) or to oxytocin
release.

At the time of deepest anestrus, i.e. within the first 2
weeks of lactation sows nurse their young in intervals of well
below 40 min at one week to more than 60 min at 18 weeks of
lactation (Niwa et al., 1951). At midlactation the interval is
about 45 min (Ellendorff et al., 1982). Each time the act of
milk letdown is preceded by a number of intense behavioural
interactions between sow and piglets during which sensory information is sent from the mammary gland to the brain and hypothalamus. Within the magnocellular system of the hypothalamus
a burst of neuronal firing heralds the periodic release of
oxytocin which then leads to milk letdown in the sow. Thus,
the wider the suckling interval, the less frequent is the

150

suckling-induced stimulation of the brain. This could suggest, that central pathways inhibitory to the release of GnRH, may be activated as part of the events leading to milk ejection. Even though this hypothesis seems very attractive, evidence for its support is lacking at present.

An alternative hypothesis could be postulated: Oxytocin which is discharged into circulation immediately prior to milk letdown (Ellendorff et al., 1982) is to be held respons-

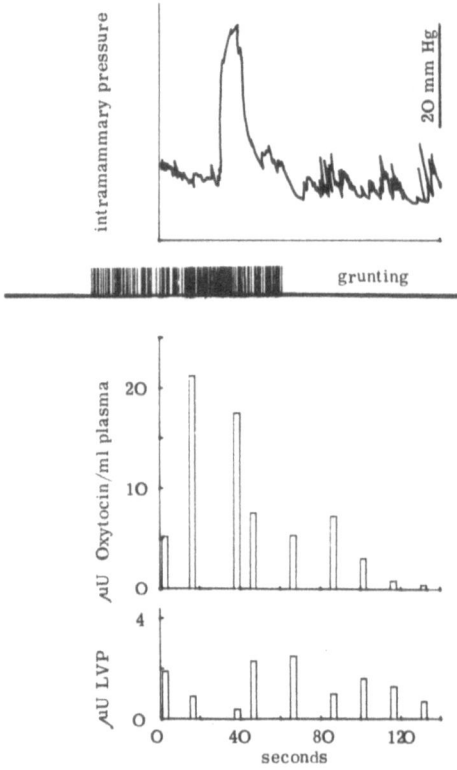

Fig. 2. Relationship between milk ejection, grunting frequency, oxytocin and vasopressin release during an individual milk ejection. This pattern is repeated approx. every 45 min at midlactation. At shorter intervals during initial and at larger intervals during later lactation.
(From Ellendorff et al., 1982).

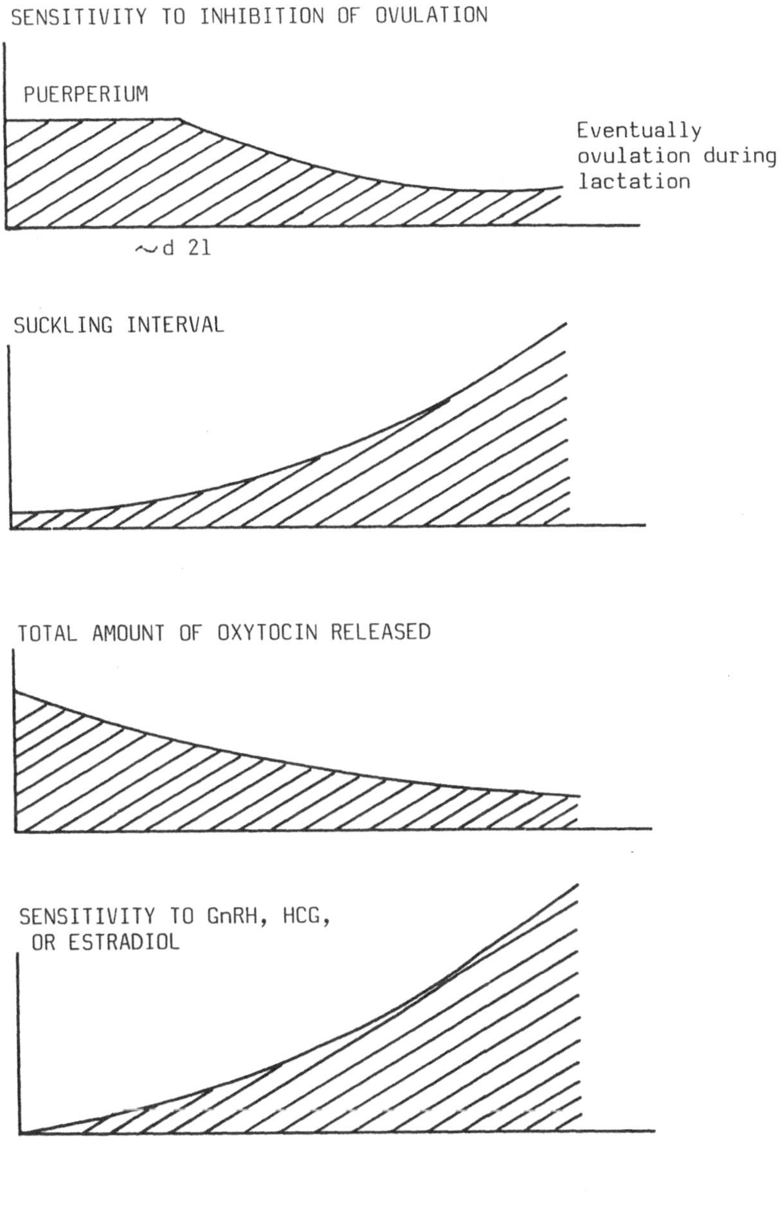

Fig. 3. Model-hypothesis of events associated with escape from lactational anestrus in sows

ible for lactational anestrus. For once, exposure frequency
to oxytocin diminishes as lactation advances and the amplitude of discharge is also smaller in later stages of lactation (> 14 days) when compared to the initial phase (Forsling
et al., 1979) and is thus inversely related to changes in
sensitivity of the pituitary-ovarian-axis. Whether or not
this hypothesis can be verified will need experimental proof.
Our own studies have sofar shown ambigous results.In one experiment oxytocin inhibited return to cyclicity,in another it
did not.

REFERENCES

Ash, R.W. and Heap, R.B. 1975. Oestrogen, progesterone and
    corticosteroid concentrations in peripheral plasma of
    sows during pregnancy, parturition, lactation and after
    weaning. J. Endocr. 64, 141-154.
Bevers, M.M., Willemse, A.H. and Kruip, Th.A.M. 1978.
    Plasma prolactin levels in the sow during lactation and
    the postweaning period as measured by radioimmunoassay.
    Biol. Reprod. 19, 628-634.
Bevers, M.M., Willemse, A.H., Kruip, Th.A.M. and van de Wiel,
    D.F.M. 1981. Prolactin levels and the LH-response to
    synthetic LH-RH in the lactating sow. Anim. Reprod. Sci.
    4, 155-163.
Britt, J.H. and Levis, D.G. 1982. Effect of altering suck-
    ling intervals of early-weaned pigs on rebreeding per-
    formance of sows. Theriogenology 18, 201-207.
Cox, N.M. and Britt, J.H. 1982. Pulsatile administration of
    gonadotropin releasing hormone to lactating sows: Endo-
    crine changes associated with induction of fertile
    estrus. Biol. Reprod. 27, 1126-1137.
Dyck, G.W., Palmer, W.M. and Simaraks, S. 1979. Postweaning
    plasma concentrations of luteinizing hormone and estro-
    gens in sows: Effects of treatment with pregnant mare's
    serum gonadotropin or estradiol-17β plus progesterone.
    Can. J. Anim. Sci. 59, 159-166.
Ellendorff, F., Forsling, M.L. and Poulain, D.A. 1982. The
    milk ejection reflex in the pig. J.Physiol. 333, 577-594.
Elsaesser, F. and Parvizi, N. 1980. Partial recovery of the
    stimulatory oestrogen feedback action on LH release
    during late lactation in the pig. J. Reprod. Fert. 59,
    63-67.
Forsling, M.L., Taverne, M.A.M., Parvizi, N., Elsaesser, F.,
    Smidt, D. and Ellendorff, F. 1979. Plasma oxytocin and
    steroid concentrations during late pregnancy, parturi-
    tion and lactation in the miniature pig. J. Endocr. 82,
    61-69.
Hodson, Jr., H.H., Hausler, C.L., Snyder, D.H., Wilkens, M.A.
    and Arthur, R.D. 1981. Effect of gonadotropin dose and
    postpartum status on induced ovulation and pregnancy in
    lactating sows. J. Anim. Sci. 52, 688-695.

Martinat-Botte, F. 1975. Induction of gestation during lactation in the sow. Ann. Biol. Anim. Bioch. Biophys. 15, 369-474.

Niwa, T., Ito,S., Yokoyama, H., Otsuka, M. 1951. Studies on milk secretion in the sow. I. On nursing habits, milk yield, milk composition, etc. Bull. natn. Inst. agric. Sci., Tokyo, Ser G 1, 135-150.

Parvizi, N., Elsaesser, F., Smidt,D. and Ellendorff, F. 1976. Plasma luteinizing hormone and progesterone in the adult female pig during the oestrous cycle, late pregnancy and lactation, and after ovariectomy and pentobarbitone treatment. J. Endocr. 69, 193-203.

Smidt, D., Scheven, B., Steinbach, J. 1965. Der Einfluß der Laktation auf die Geschlechtsfunktion bei Sauen. Züchtungskunde 37, 23-36.

Stephenson, J.S., Cox, N.M. and Britt, J.H. 1981. Role of the ovary in controlling luteinizing hormone, follicle stimulating hormone, and prolactin secretion during and after lactation in pigs. Biol. Reprod. 24, 341-353.

van Landeghem,A.A.J. and van de Wiel, D.F.M. 1978. Radioimmunoassay for porcine prolactin: Plasma levels during lactation, suckling and weaning and after TRH administration. Acta Endocr. 88, 653-667.

RELEVANCE OF PROLACTIN TO LACTATIONAL AND POST-WEANING ANOESTRUS
IN THE PIG

D.F.M. van de Wiel*, P. Booman*, A.H. Willemse** and M.M. Bevers**

* Research Institute for Animal Production "Schoonoord", P.O. Box 501,
3700 AM Zeist, The Netherlands
** Clinic for Veterinary Obstetrics, Reproduction and Artificial
Insemination, State University of Utrecht, P.O. Box 80151,
3508 TD Utrecht, The Netherlands

ABSTRACT

The relevance of prolactin (PRL) to lactational and post-weaning anoestrus in the pig has been investigated. Evidence has been shown that a
negative relationship exists between PRL and LH during lactation and at
weaning, within animals. However, long intervals between weaning and oestrus
are not caused by a persistent hyperprolactinaemia after weaning. Suppression of PRL during lactation with CB_{154} causes a significant increase in LH
concentrations, although the level is still lower as compared to post-
weaning concentrations. Increasing PRL concentrations after weaning to
physiological levels by PRL infusion causes a decrease in LH concentrations,
but the level is still higher as compared to pre-weaning concentrations.
From these results we conclude that probably PRL can partially suppress
LH during lactation and after weaning, but other factors such as direct
neural stimuli related to suckling may also suppress LH.

Although there is some evidence that PRL can act directly on the ovary
in the pig, we conclude from literature data and from experiments with LHRH
stimulation of the pituitary and oestradiol-benzoate stimulation of the
hypothalamo-pituitary axis, that the observed LH suppressive effect is
exerted at the hypothalamic level.

INTRODUCTION

It has been estimated that in the Netherlands, post-weaning anoestrus
in the pig causes on average a loss of 10 days per litter produced. With
1 million of sows producing 1.9 litters per sow annually and a calculated
cost of Dfl. 4.- per sow per day, the total loss amounts to approximately
Dfl. 80 million per year (NRLO-report, 1979). This has prompted our laboratories to investigate possible causes of post-weaning anoestrus, especially in the area of reproductive endocrinology. Because of its impredictabil-
ity in individual animals, post-weaning anoestrus has mainly been studied
in relation to lactational anoestrus, which is a highly predictable phenomenon and which may be caused by the same underlying endocrine mechanism
as post-weaning anoestrus.

Initially our hypothesis has been, that persistent hyperprolactinaemia
after weaning may be the cause of delayed (later than day 10 post-weaning)
return to oestrus. From studies by our laboratories (Van Landeghem and

Van de Wiel, 1978; Bevers et al., 1978) and others (Stevenson et al., 1981; Benjaminsen et al., 1981; Kirkwood et al., 1984) it has become clear that prolactin levels in the pig are elevated during lactation, and decrease to basal levels within a few hours after weaning. However, in animals with a delayed return to oestrus, prolactin levels decrease in the same way as compared to animals with a normal return to oestrus (Van de Wiel et al., 1979; Benjaminsen, 1981). Kirkwood et al. (1984) have shown some evidence that a slow rate of decrease of plasma prolactin level after weaning may be related to post-weaning anoestrus, but this may also be explained on the basis of the relatively high prolactin levels at the earlier stages of lactation.

Evidence that prolactin can be a potent inhibitor of ovarian function has been obtained in several species (McNeilly et al., 1982). In the rat, an inhibitory effect of hyperprolactinaemia on LH secretion has been demonstrated at the level of the hypothalamus (Carter et al., 1983). We have investigated the hypothesis that hyperprolactinaemia in the pig may cause a decrease in LH secretion, which in turn may be responsible for lactational anoestrus. Evidence that plasma LH levels are suppressed during lactation and increase after weaning has been reported by several groups (Parvizi et al., 1976; Van de Wiel et al., 1978; 1979; Booman and Van de Wiel, 1980; Booman et al., 1982; Cox and Britt, 1982a; Edwards and Foxcroft, 1983) although no such increase has been observed by others (Dyck et al., 1979; Kirkwood et al., 1984). In addition, there exists some evidence that the pre-weaning level of LH and the magnitude of the LH rise after weaning may be related to the time interval between weaning and first ovulation (Booman et al., 1983). In a series of experiments we have investigated 1) a possible relationship between plasma prolactin- and LH-levels during lactation, 2) the effect of prolactin infusion on LH after weaning and of prolactin inhibition on LH during lactation and 3) possible sites of action of prolactin.

RELATIONSHIP BETWEEN PLASMA LEVELS OF PROLACTIN AND LH

Prolactin and LH during suckling

In an experiment with 5 primiparous sows (Dutch Landrace) suckling 6-12 piglets, bloodsamples were taken frequently (1 minute intervals) and suckling behaviour was recorded simultaneously at 6 - 34 days after parturition (p.p.). Suckling periods were subdivided in 1) stimulation of the udder by the piglets, 2) milk ejection, and 3) stimulation of the

156

udder after milk ejection. Prolactin was measured by a validated RIA procedure (Van Landeghem and Van de Wiel, 1978). Mean plasma levels are given
in Table I.

TABLE I.   Mean concentrations of plasma prolactin at different stages
           of suckling (Booman and Van de Wiel, 1980).

	Prolactin (ng/ml)	
Stage of suckling behaviour	mean level	(± S.D., n = 20)
Start of udder stimulation	16.9[a,b]	(± 5.9)
Start of milk ejection	18.5	(± 6.1)
Start of after-stimulation	20.9[a]	(± 6.8)
End of after-stimulation	21.2[b]	(± 6.3)

Figures with the same index are significantly different
(a indicates p < 0.10; b indicates p < 0.05, Students t-test).

From the data in Table I it can be seen that indeed prolactin levels do increase during the suckling period, although the basal level is already
very high as compared to basal levels in non-lactating animals.

   From the measurement of LH in the same samples by a validated RIA procedure (Van de Wiel et al., 1981) it appeared that on two occasions where
a marked decrease in prolactin level was observed, this decrease was accompanied by a simultaneous rise in LH (one example is shown in Fig.I).

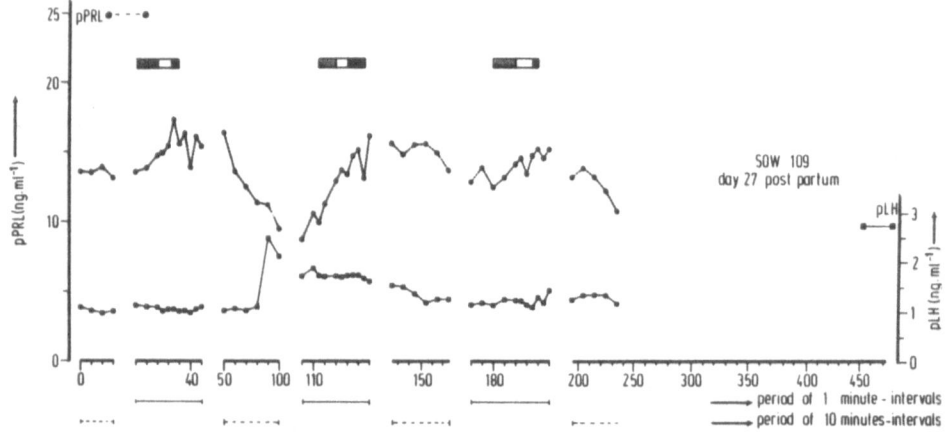

Fig.I   Decrease of prolactin and simultaneous increase of LH on one
occasion during lactation in the sow. ▬▭ suckling period, consisting
of pre-stimulation of udder ▬ , milk ejection ▭ and after-stimula-
tion ▬ .

In another experiment with 18 primiparous crossbred Landrace sows suckling 6 - 12 piglets, frequent bloodsampling (10 min. intervals) was done at 14 and 21 days p.p. All samples were analyzed for LH and prolactin. Analysis of (co)variance showed a negative (but non-significant) correlation within animals between mean prolactin levels and either basal LH level (r=-0.09) or mean LH level (r=-0.19). Between animals however these correlations were positive, although also non-significant (0.17 and 0.14, respectively). This apparent discrepancy can possibly be explained by the assumption that, if prolactin has an inhibitory effect on LH secretion, animals with high levels of LH need also high levels of prolactin in order to suppress LH (Booman et al., 1983).

Prolactin and LH before and after weaning

Seven pluriparous crossbred (Duroc x Landrace) sows suckling 8-13 piglets were bled frequently (10 min.intervals during 6 hour periods) immediately before and one day after weaning at day 21 p.p. Mean prolactin levels decreased from 33.8 ± 13.3 ng/ml before weaning to 2.5 ± 1.2 ng/ml after weaning. Mean LH levels before and after weaning were 1.22 ± 0.62 and 1.75 ± 0.86 ng/ml, respectively. The increase of LH after weaning (0.53 ng/ml) was highly significant (p < 0.01, Students t-test). Basal LH levels increased significantly (p < 0.01) from 1.09 ± 0.58 to 1.48 ± 0.80 ng/ml, and LH pulse frequency (number of pulses per 6 hour period) from 1.57 ± 1.62 to 4.00 ± 2.08 (p < 0.05) (Booman et al., 1982).

From these observations we conclude that there are strong indications for a negative relationship between plasma levels of prolactin and LH within animals. However, in the experiments described it is not possible to distinguish negative effects on LH caused by prolactin per se, from those caused by direct neural stimuli due to suckling.

SUPPRESSION OF LH DURING LACTATION: DIFFERENTIATION BETWEEN THE EFFECTS OF HYPERPROLACTINAEMIA AND THE SUCKLING STIMULUS

In order to investigate if high plasma prolactin levels are solely responsible for the suppression of LH during lactation, two experiments have been done. The idea behind these experiments was to separate the effects of high levels of circulating prolactin from the effects of neural stimuli related to suckling. One approach has been to take away the suckling stimulus (=weaning), while at the same time artificially maintaining high prolactin

158

levels by infusion of purified prolactin. The other approach has been to artificially suppress high prolactin levels with CB_{154}, while at the same time maintaining the suckling stimulus.

Prolactin infusion after weaning

Purified porcine prolactin (pPRL) was infused during a 24 hour period immediately after weaning. Biological activity of prolactin was verified by the pigeon crop sac bioassay (Nicoll, 1967). Crossreactivity in our RIA systems for pLH and pFSH was less than 0.01 %. ACTH activity was 370 ng/mg by RIA and 136 ng/mg (corresponding to 13.6 mIU) by bioassay (Goverde, 1981).

Infusion rate was 500 μg of pPRL per hour, resulting in plasma prolactin levels which were within the normal range seen during lactation. Fourteen pluriparous crossbred (Duroc x Landrace) sows were infused immediately after weaning at day 21 ± 1 p.p. either with prolactin (n=7, PRL group) or with saline (n=7, saline group). Four sows from which the piglets were not weaned, served as additional controls (control group). The time schedule of blood-sampling, piglet removal and prolactin or saline infusion can be seen in Figure 2.

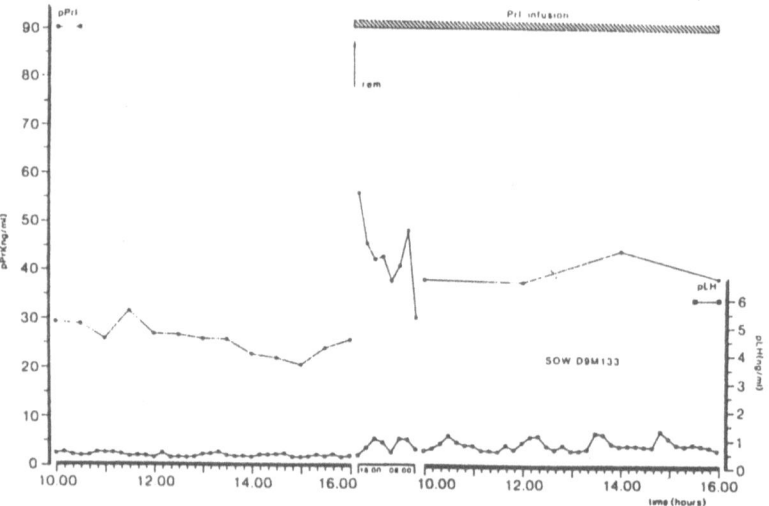

Fig.2 Effect of prolactin infusion on LH after weaning (rem = removal of piglets)

Analysis of the data by (co)variance and Students t-test showed that after weaning mean LH level, basal LH level and LH pulsfrequency were significantly lower in the prolactin group than in the saline group, but they were still significantly higher than in the non-weaned control group ($p < 0.05$).

Prolactin had no influence on the mean amplitude of the LH pulses. From these results we conclude that suppression of LH during lactation probably is partly caused by hyperprolactinaemia, but partly also by neural factors related to suckling.

The amount of ACTH infused together with prolactin (6.8 mIU per hour) was very low, and it seems unlikely that this may have caused the LH suppression. However, it cannot completely be excluded that other unknown factors in the prolactin preparation may have had an effect on LH (Booman and Van de Wiel, 1980; Booman et al., 1982).

Suppression of prolactin by CB_{154} during lactation

Ten primiparous sows (Large White breed) were given either 10 mg bromocryptin (CB_{154}) orally twice a day, from day 14 to day 22 p.p. (n=5, CB_{154} group) or a placebo (n=5, control group). Piglets were weaned on day 22 p.p. and sows were bled frequently (10 min. intervals) on days 12, 16, 20 and 24 p.p. during 6 hour periods.

CB_{154} treatment resulted in a significant decrease in mean prolactin levels and a concomitant significant increase in LH levels. After weaning a further significant increase in LH was observed, although the levels of prolactin did not further decrease. Suckling frequency or milk production were not affected by CB_{154} treatment, as shown by the normal weight gain of the piglets (Bevers et al., 1983).

We conclude that high levels of circulating prolactin partially account for LH suppression during lactation, but other factors such as direct neural inputs due to suckling, also play an important role.

SUPPRESSION OF OVARIAN CYCLICITY BY PROLACTIN: SITE OF ACTION

There are three sites where prolactin may directly interfere with the normal process of ovarian cyclicity: the ovary, the pituitary and the hypothalamus.

Ovary

A direct inhibitory action of prolactin on the ovaries has been demonstrated in the pig by in vitro incubation of granulosa cells from small follicles with prolactin, resulting in a decrease in progesterone secretion (Veldhuis and Hammond, 1980; Veldhuis et al., 1981). In an in vivo experiment we have obtained evidence of decreased responsiveness of the ovaries

160

to gonadotrophic stimulation early in lactation. Single injections of PG 60C
(400 IU PMSG + 200 IU HCG) were given to 6 lactating sows either at days
5 - 10 p.p. (n=3) or at days 16 - 21 p.p. (n=3). Early in lactation, PG 600
injection did not result in any sustained elevation of oestradiol-17β (E_2),
whereas later in lactation a marked E_2 response in 2 out of 3 animals was
observed (Van de Wiel et al., unpublished results). At 4 weeks p.p. pulsatile administration of LHRH could induce oestrus and ovulation in the
presence of high prolactin levels (Cox and Britt, 1982b). It may be that
inhibition of ovarian responsiveness early in lactation is due to high
prolactin levels, but other factors may play a role as well.

Injection of PG 600 at day 21 after weaning in 7 crossbred Landrace
sows with postweaning anoestrus (weaning at day 28 p.p.) resulted in a
physiological E_2 response followed by LH surge, oestrus and ovulation in
only 3 animals. In 3 other animals the E_2 response was low (< 18 pg/ml) and
no LH surge, oestrus or ovulation were observed (Fig. 3). In the one remaining animal there was no response at all (Booman et al., 1983).

We conclude that the early stages of lactational anoestrus and some
cases of post-weaning anoestrus can be caused by blocked ovarian function.
Early in lactation this blockade may be the result of high prolactin levels
However, both in lactational and post-weaning anoestrus blocked ovarian
function may be caused by a lack of appropriate previous stimulation by
LH and/or FSH.

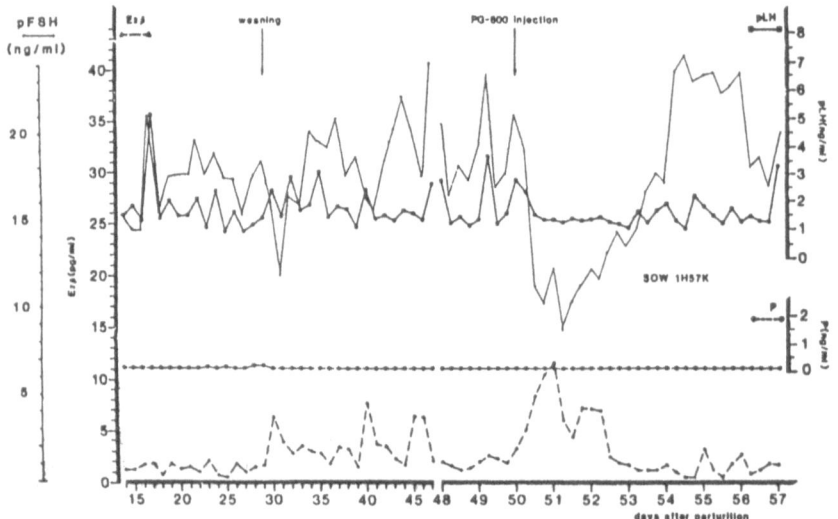

Fig.3 Insufficient E_2 response and absence of LH surge, oestrus and
ovulation after PG 600 injection in an anoestrous sow at 3 weeks post-
weaning (P = progesterone).

Pituitary

The responsiveness of the pituitary to LHRH has been tested by injection of single intravenous doses of 25 µg LHRH in the 1st, 2nd or 3rd week of lactation, in 3 primiparous Large White sows, suckling 4-12 piglets.

LH responses increased as lactation progressed, but this effect could not be related to a decrease in circulating prolactin levels. This finding was confirmed in experiments where prolactin levels were lowered either by CB_{154} or by temporary removal of the piglets. None of these treatments did influence the LH response after LHRH injection (Bevers et al., 1981).

Hypothalamus

In the sow, LHRH concentrations in hypothalamic tissue increase after weaning, which points to a suppression of LHRH synthesis during lactation (Cox and Britt, 1982a). This is in line with the observation that hourly administration of 1 µg doses of LHRH at days 28-32 p.p. can induce oestrus and ovulation during lactation (Cox and Britt, 1982b). Evidence that hypo-thalamic-pituitary function may be blocked during lactation comes from ex-periments where E_2-benzoate (E_2B) has been injected into lactating sows. Early in lactation (day 5 p.p.) E_2B injection did not result in a positive feedback on LH, but on day 35 p.p. there was a small but significant posi-tive LH response (Elsaesser and Parvizi, 1980).

In weaned sows, the positive LH response after E_2B was bigger when the sows were weaned at 3 weeks as compared to 5 weeks, thus pointing to a stronger inhibition of the hypothalamus-pituitary system at the earlier stages p.p. (Edwards and Foxcroft, 1983).

If prolactin is involved in the inhibition of hypothalamic function in the pig, it probably will not act by increasing the sensitivity of the hypo-thalamus to the negative feedback of E_2 (McNeilly, 1980). Ovariectomy during lactation does not result in a rise of circulating LH (Parvizi et al., 1976; Stevenson et al., 1981). However, it can not be ruled out that a post castration rise of LH during lactation is suppressed by the inhibiting effect of the suckling stimulus.

A possible mechanism for LH suppression by prolactin is related to the activation by prolactin of the dopaminergic terminals in the external layer of the median eminence, as observed by Hökfelt and Fuxe (1972). Dopaminergic pathways have been shown to be involved in the control of the release of LH (Weiner and Ganong, 1978).

162

CONCLUSIONS

We conclude that LH level and pulse frequency are reduced and prolactin levels are elevated during lactation. With lactation progressing, suppression of ovarian function becomes less, not only at the ovarian level but also at the pituitary and hypothalamic level, while at the same time prolactin levels gradually decrease. After weaning, which normally occurs around day 35 p.p., prolactin levels fall sharply, and LH basal level and pulse frequency show a significant increase. From the LH rise after CB_{154} treatment during lactation, and the LH suppression due to prolactin infusion after weaning, we conclude that prolactin has a suppressive effect on LH.

From literature data it seems likely that LH suppression is caused by inhibition of LHRH release, which points at a blockade at the hypothalamic level or higher. Prolactin can only partially account for suppression of LH during lactation, and direct neural inputs related to suckling probably also play an important role in LH suppression.

REFERENCES

Benjaminsen, E., 1981. Plasma prolactin in the sow with emphasis on variation in resumption of ovarian activity after weaning. Acta Vet. Scand., 22, 67-77.

Bevers, M.M., Willemse, A.H. and Kruip, Th.A.M., 1978. Plasma prolactin levels in the sow during lactation and the postweaning period as measured by radioimmunoassay. Biol.Reprod., 19, 628-634.

Bevers, M.M., Willemse, A.H., Kruip, Th.A.M. and Van de Wiel, D.F.M., 1981. Prolactin levels and the LH-response to synthetic LH-RH in the lactating sow. Anim.Reprod.Sci., 4, 155-163.

Bevers, M.M., Willemse, A.H. and Kruip. Th.A.M., 1983. The effect of bromocryptine on luteinizing hormone levels in the lactating sow: evidence for a suppressive action by prolactin and the suckling stimulus. Acta Endocrinol., 104, 261-265.

Booman, P. and Van de Wiel, D.F.M., 1980. Lactational anoestrus in the pig: possible relationship with hyperprolactinaemia. IVO-report B-157, Zeist, The Netherlands.

Booman, P., Van de Wiel, D.F.M. and Jansen, A.A.M., 1982. Effect of exogenous prolactin on peripheral luteinizing hormone levels in the sow after weaning of the piglets. IVO-report B-200, Zeist, The Netherlands.

Booman, P., Van de Wiel, D.F.M. and Jansen, A.A.M., 1983. Postweaning anoestrus in sows: LH-patterns, and effect of gonadotrophin administration and boar introduction. IVO-report B-217, Zeist, The Netherlands

Carter, D.A., Lakhani, S. and Whitehead, S.A., 1983. Characterization of the inhibitory effects of hyperprolactinaemia on the mechanism controlling LH secretion in chronically ovariectomized rats. J.Reprod.Fert., 69, 57-64.

Cox, N.M. and Britt, J.H., 1982a. Relationship between endogenous gonadotropins, and follicular development after weaning in sows. Biol.Reprod., 27, 70-78.

Cox, N.M. and Britt, J.H., 1982b. Pulsatile administration of gonadotropin releasing hormone to lactating sows: endocrine changes associated with induction of fertile estrus. Biol.Reprod., 27, 1126-1137.

Dyck, G.W., Palmer, W.M. and Simaraks, S., 1979. Postweaning plasma concentrations of luteinizing hormone and estrogens in sows: effect of treatment with pregnant mare's serum gonadotropin or estradiol-17β plus progesterone. Can.J.Anim.Sci., 59, 159-166.

Edwards, S. and Foxcroft, G.R., 1983. Endocrine changes in sows weaned at two stages of lactation. J.Reprod.Fert., 67, 161-172.

Elsaesser, F. and Parvizi, N., 1980. Partial recovery of the stimulatory oestrogen feedback action on LH release during late lactation in the pig. J.Reprod.Fert., 59, 63-67.

Goverde, H., 1981. Bioactive versus immunoreactive adrenocorticotrophin in human blood. Thesis, University of Nijmegen, The Netherlands.

Hökfelt, T. and Fuxe, K., 1972. Effects of prolactin and ergot alkaloids on the tubero-infundibular dopamine (DA) neurons. Neuroendocrinology, 9, 100-122.

Kirkwood, R.N., Lapwood, K.R., Smith, W.C. and Anderson, I.L., 1984. Plasma concentrations of LH, prolactin, oestradiol-17β and progesterone in sows weaned after lactation for 10 or 35 days. J.Reprod.Fert., 70, 95-102.

McNeilly, A.S., Glasier, A., Jonassen, J. and Howie, P.W., 1982. Evidence for a direct inhibition of ovarian function. J.Reprod.Fert., 65, 559-569.

Nicoll, C.S., 1967. Bioassay of prolactin. Analysis of the pigeon crop-sac response to local prolactin injection by an objective and quantitative method. Endocrinology, 80, 641-655.

NRLO report on "Fertility of the female pig", 1979. IVO-report B-133, Zeist, The Netherlands.

Parvizi, N., Elsaesser, F., Smidt, D. and Ellendorff, F., 1976. Plasma luteinizing hormone and progesterone in the adult female pig during the oestrous cycle, late pregnancy and lactation, and after ovariectomy and pentobarbitone treatment. J.Endocr., 69, 193-203.

Stevenson, J.S. and Britt, J.H., 1980. Luteinizing hormone, total estrogens and progesterone secretion during lactation and after weaning in sows. Theriogenology, 14, 453-462.

Stevenson, J.S., Cox, N.M. and Britt, J.H., 1981. Role of the ovary in controlling luteinizing hormone, follicle stimulating hormone and prolactin secretion during and after lactation in pigs. Biol.Reprod., 24, 341-353.

Van Landeghem, A.A.J. and Van de Wiel, D.F.M., 1978. Radioimmunoassay for porcine prolactin: plasma levels during lactation, suckling and weaning and after TRH administration. Acta Endocr., 88, 656-667.

Van de Wiel, D.F.M., Van Landeghem, A.A.J., Bevers, M.M. and Willemse, A.H., 1978. Endocrine control of ovarian function during lactation in the sow. In "Comparative Endocrinology" (Ed. P.J. Gaillard and H.H.Boer). (Elsevier/North-Holland Biomedical Press, Amsterdam/New York). pp.378.

Van de Wiel, D.F.M., Van Landeghem, A.A.J., Willemse, A.H. and Bevers, M.M., 1979. Endocrine control of ovarian function after weaning in the domestic sow. J.Endocr., 80, 69 P.

Van de Wiel, D.F.M., Erkens, J., Koops, W., Vos, E. and Van Landeghem, A.A.J., 1981. Periestrous and midluteal time courses of circulating LH, FSH, prolactin, estradiol-17β and progesterone in the domestic pig. Biol. Reprod., 24, 223-233.

Veldhuis, J.D. and Hammond, J.M., 1980. Oestrogens regulate divergent effects of prolactin in the ovary. Nature, Lond., 284, 262-264.

Veldhuis, J.D., Klase, P. and Hammond, J.M., 1980. Divergent effects of prolactin upon steroidogenesis by porcine granulosa cells in vitro: influence of cytodifferentiation. Endocrinology, 107, 42-46.

Weiner, R.I. and Ganong, W.F., 1978. Role of brain monoamines and histamine
    in regulation of anterior pituitary secretion. Physiol.Rev., 58, 905-
    976.

EFFECTS OF BROMOCRIPTINE TREATMENT DURING

LACTATIONAL ANESTRUS IN PIGS

M. Mattioli, E. Seren

Istituto di Fisiologia Veterinaria
Via Belmeloro,8/2,40126 Bologna,Italy

ABSTRACT

The Authors described the effect of bromocriptine treatment in lactating sows. Bromocriptine (10 mg) were administered per os twice daily
during the five days preceding weaning (21 days post-partum). The treatment
could reduce the percentage of animals which did not return to estrus
within 10 days of weaning and, among animals returning within 10 days, it
reduced the weaning to estrus interval of about one day. In one piggery,
where the post-partum reproductive efficiency was optimal, the treatment
did not affect the percentage of sows returning to estrus within 10 days of
weaning. Trials carried out in a high (April-May) and a low (July-August)
fertility period revealed that the effectiveness of the treatment was not
related to the season. The bromocriptine treatment during the last stages
of lactation did not affect the milk production as evaluated by the litter
body weight gain and by the suckling behaviour.

The treatment reduced plasma PRL and maintained it at levels similar
to those observed after weaning. Estradiol and LH profiles, however, were
not different in control and treated sows. The effect of bromocriptine
treatment, i.e. PRL withdrawal, on the sensitivity to GnRH of the pituitary
gland was also investigated. A single injection of GnRH (500 µg) could
induce a more marked increase of plasma LH in control than in treated sows.
Finally the effect of bromocriptine treatment during the lactational
anestrus on the stimulatory estrogen feed back of LH was evaluated. Data
obtained indicated that bromocriptine enhanced, although not significantly,
the stimulatory estrogen feed-back.

INTRODUCTION

Lactation in the domestic pig is characterized by anestrus and suppres-

sion of ovarian activity (Burger, 1952) and by high levels of plasma PRL

(van Landeghem and van de Viel, 1978; Bevers et al., 1978; Stevenson et al.

1981; Duszal and Krzimowsky, 1981) and low levels of progesterone and estro-

gens (Edqvist et al., 1974; Ash and Heap, 1975; Parvizi et al., 1976;

Stevenson et al., 1981). Weaning results in a prompt resumption of ovarian

activity which terminates in estrus 5-10 days later. Immediately after

weaning a drop in PRL levels occurs and a progressive rise in LH concentration takes place (Stevenson and Britt, 1980; Stevenson et al., 1981). The mechanisms controlling these events are unclear but have generally been associated with the suckling stimulus (Lauerdale et al.,1965; Stevenson and Britt, 1981) although it is unknown how this stimulus can affect ovarian activity. Since hyperprolactinemia is closely coupled with the suckling stimulus it might be hypothesized that high PRL levels are responsible for the suppression of ovarian activity during lactation. This inhibition might take place at different levels such as hypothalamus hypophysis or ovary. An antigonadal role of prolactin has been described in sheep(Kann et al.,1976) rats (Lu et al., 1976) and humans (Roland et al.,1970).

The purpose of this study was to evaluate if the reduction of plasma prolactin induced by means of bromocriptine treatment during the last stages of lactation may accelerate the resumption of ovarian activity and to investigate how PRL is involved, if it is, in the suppression of the ovarian function during lactation. The research was divided into two parts: the first consisted of some field trials conducted to evaluate the repro-ductive findings in bromocriptine treated sows while the second part was designed to investigate the hormonal changes induced by the treatment.

REPRODUCTIVE FINDINGS AFTER BROMOCRIPTINE TREATMENT

The first field trial was conducted in an intensive piggery during the months of April–June 1983. Only pluriparous sows were used in order to avoid the irregular and long weaning to estrus intervals which are frequen-tly observed in primiparous post–partum sows. The bromocriptine was administered per os (10 mg twice a day) during the 5 days preceding weaning, which occurred 21 days post–partum, in order to produce an endocrine situation similar to that observed after weaning at least as far as PRL is concerned. If PRL were the only suppressor of ovarian activity, its inhibi-ted secretion during the last 5 days of lactation should have resulted in animals returning to estrus around about the day of weaning since estrus and ovulation generally occur only about 5 days after weaning i.e. after PRL withdrawal. In contrast to this hypothesis the treatment only reduced the number of animals which returned to estrus more than 10 days after weaning and, among the sows returning to estrus within 10 days of weaning, it significantly reduced the weaning to estrus interval (Table 1). When the same treatment was administered to sows weaned 4 weeks post-partum it neither reduced weaning to estrus interval nor the number of animals returning to estrus 10 days after weaning (Table 1).

TABLE 1   Reproductive findings in bromocriptine treated and control sows. 168 pluriparous sows were used during the months April–June. The animals were divided into two equal groups which were weaned 3 and 4 weeks post-partum respectively. Half of each group was treated with bromocriptine, 10 mg per os twice a day, for the 5 days preceding weaning and half was kept as control. After weaning estrus detection was carried out daily.

weaning		W–E interval($^\circ$)	% of sows not in estrus by day 10
3 weeks p.p.	control	5.9 ± 0.32	39
	treated	$4.7 \pm 0.18(^{\circ\circ})$	18
4 weeks p.p.	control	5.4 ± 0.36	38
	treated	5.0 ± 0.29	26

($^\circ$) W–E Interval = weaning to estrus interval evaluated among sows which returned to estrus within 10 days of weaning.

($^{\circ\circ}$) $P < 0.01$

These data indicated that, first, PRL is not the main suppressor of ovarian activity during lactation and, second, the inhibitory effect of prolactin, if any, becomes less and less effective as the lactation proceeds. Taking these data into consideration all the subsequent experiments were carried out using sows weaned 3 weeks post-partum. The results obtained from two more experiments carried out in two different units, in the same season, but one year later, only in part confirmed previous results. In fact in one unit (B), where the post weaning reproductive efficiency could be considered as optimal, the bromocriptine treatment only reduced, and not significantly, the weaning to estrus interval (Table 2). On the other hand in the other unit (A), where the reproductive efficiency post weaning was not so good, results similar to those of the first trial were recorded. The two units were of good health status and no outbreaks of disease occurred

TABLE 2   Reproductive findings in bromocriptine treated and
control sows, weaned 3 weeks post-partum, belonging to two
different commercial herds (in brackets the number of sows).

		W–E interval	% of sows not in estrus by day 10
Unit A	control (70)	5.39 ± 0.31	52
	treated (60)	5.07 ± 0.26	28
Unit B	control (40)	5.48 ± 0.34	6
	treated (40)	4.72 ± 0.22	10

during the experimental period so that no evident causes accounting for the different reproductive efficiency of the units could be found. Therefore the high degree of variability of reproductive efficiency which usually occurs among piggeries seems to markedly affect the effectiveness of the bromocriptine treatment.

At present there is evidence that the resumption of ovarian activity after weaning is markedly affected by the season. Corteel et al. (1964), Love (1978), Tarocco et al. (1979) and Hurtgen et al. (1980) observed an increased weaning to estrus interval during the months of July to September. In this period the general reproductive efficiency falls. We tried the bromocriptine treatment in this period in order to evaluate if it might overcome or reduce the problem of the summer delayed returns to estrus after weaning. Such a treatment might also have been appropriate since a slight hyperprolactinemia occurs in the sow during the summer (Ravault et al., 1982). However the experiment carried out on 30 sows in a high fertility period (April-May) and 30 sows in the low fertility period (July-August), did not reveal any increased effectiveness of the treatment during the summer (Table 3). Since these data refer to one unit only they can not be considered as conclusive.

EFFECT OF BROMOCRIPTINE TREATMENT ON MILK PRODUCTION

One of the principal actions of prolactin in mammals is stimulation of milk formation in hormonally prepared female breasts. As far as the pig is concerned, Taverne et al. (1982) showed that the inhibition of prolactin by means of bromocriptine treatment during the last week of pregnancy, caused a complete inhibition of udder development and impeded the onset of lactation. Smith and Wagner (1980) suppressed lactation in the pig by

TABLE 3   Effect of bromocriptine treatment during a high (April-May) and a low (July-August) period on the reproductive efficiency post-partum (in brackets the number os sows).

		W-E interval	% of sows not in estrus by day 10
April-May	control (30)	5.03 ± 0.23	12
	treated (30)	4.48 ± 0.16	5
July-August	control (30)	4.75 ± 0.25	33
	treated (30)	4.30 ± 1.13	27

administering bromocriptine shortly after parturition. In our experiments bromocriptine given during the last 5 days of lactation, i.e. between day 15 and 20 post-partum, did not affect milk production as evaluated by the litter body weight gain and by the suckling behaviour (Table 4). Similar

TABLE 4   Effect of bromocriptine treatment on lactation. Litter body weight gain and suckling behaviour (suckling length and frequency recorded by observing the sows for 6 hours daily) were assumed as indexes of milk production.

		days from weaning		
		-5	-3	-1
suckling frequency (sucks/hours)	treated	1.75 ± 0.09	1.56 ± 0.13	1.37 ± 0.06
	control	1.57 ± 0.15	1.53 ± 0.15	1.28 ± 0.12
suckling (minutes per hour)	treated	9.75 ± 1.12	9.00 ± 1.38	8.12 ± 1.23
	control	13.85 ± 2.79	9.14 ± 1.73	10.42 ± 2.10
body weight (Kg)	treated	3.96 ± 0.23	---	4.62 ± 0.23
	control	4.18 ± 0.23	---	4.97 ± 0.21

findings were reported by Bevers et al.(1983).

Therefore prolactin plays a fundamental role in udder development and

the onset of lactation but once lactation is well established milk production does not seem to need high PRL levels any longer.

ENDOCRINE CHANGES ASSOCIATED WITH BROMOCRIPTINE TREATMENT

It is generally accepted that a major factor controlling the lack of
sustained follicular development during lactation is the inadequate secretion of LH. Previous works reported suppressed LH levels during lactation
followed by an increase of the levels of this hormone after weaning (Parvizi et al., 1976; Stevenson and Britt, 1980; Stevenson et al., 1981). In
this context there is a number of works reporting that prolactin can
suppress LH secretion (Kann et al.,1977; Wuttke et al., 1980; Thorner et
al., 1980) possibly by inhibiting LH-RH secretion. Plasma prolactin was
therefore reduced in order to evaluate if an inverse relation between PRL
and LH occurs during lactation. In our experiments the bromocriptine
treatment carried out during the 5 days preceding weaning could effectively reduce PRL and maintain it at levels similar to those observed after
weaning (Fig. 1). Despite this, the levels of LH did not show any significant increase related to the inhibited prolactin secretion during lactation
and no differences between control and treated animals were observed
(Fig. 1 - 2). Frequent sampling did not reveal any change in the episodic
secretion pattern of LH (Fig. 3). LH started to rise in both control and
treated animals only after weaning doubling the levels recorded during
lactation 3 days after weaning (Fig. 3). In this stage the episodic release
of LH became more ordered with a pulse frequency of about 1/hour. Estrogen,
PRL and LH profiles during the first estrus after weaning were not different
in control and treated sows. Different results were reported by Bevers et
al. (1983) who described a slight increase in LH concentration after
bromocriptine treatment during lactation.

The pituitary gland of the lactating sows responds to GnRH by releasing
LH(Stevenson et al.,1981) and LH response to 25 µg GnRH increases from the
first to the third week of lactation (Bevers et al., 1981). Therefore the
next approach used to investigate the role of PRL and the effect of bromocriptine treatment was to test if the reduction of plasma PRL during

171

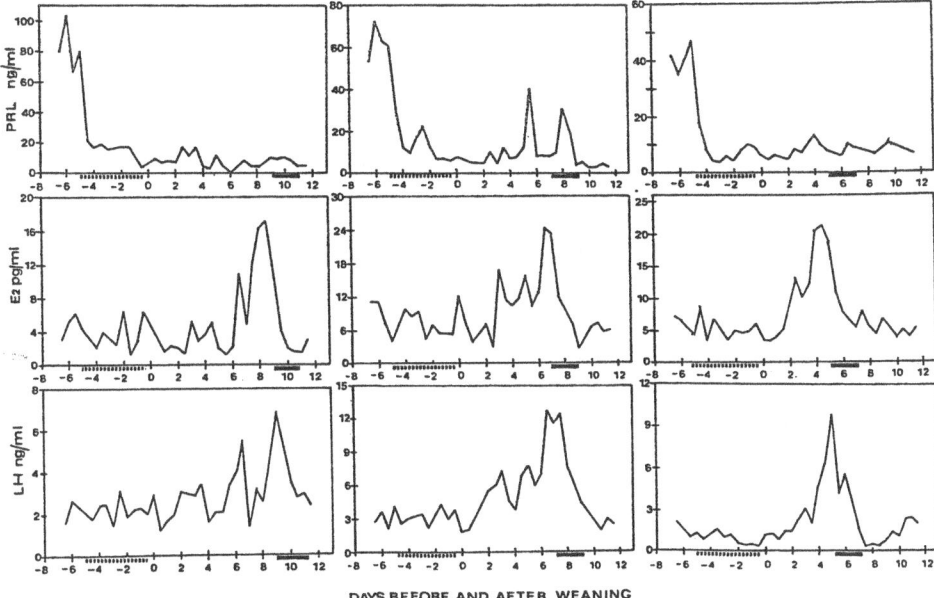

Fig. 1   Endocrine changes recorded in 3 sows treated with bromocriptine (10 mg per os twice daily) during the 5 days preceding weaning
(day 21 post-partum). Blood sampling was carried out through an indwel-
ling silastic catheter fitted in the ear vein. (▦▦▦ bromocriptine
treatment; ━━ estrus).

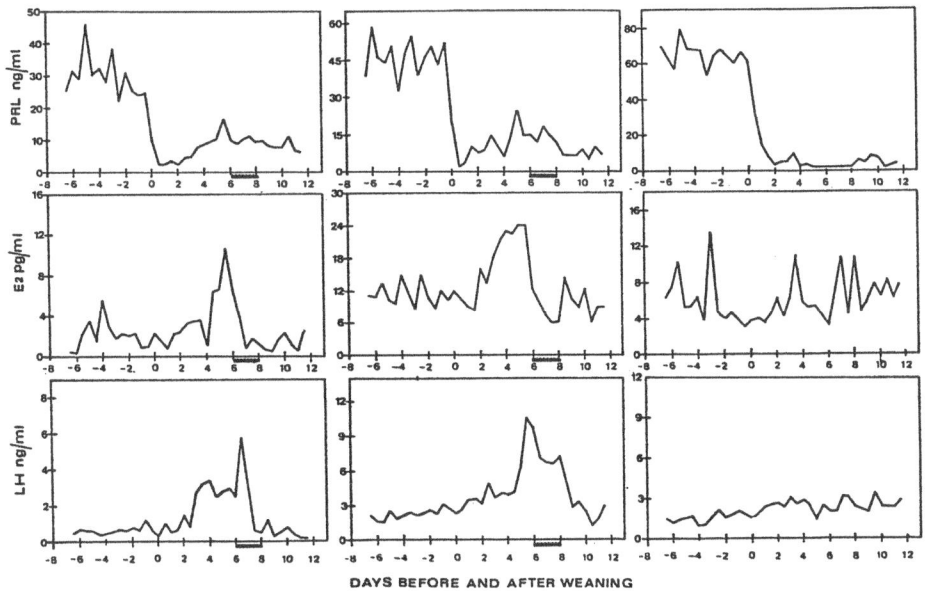

Fig. 2   Endocrine changes in 3 control sows around weaning(━━estrus)

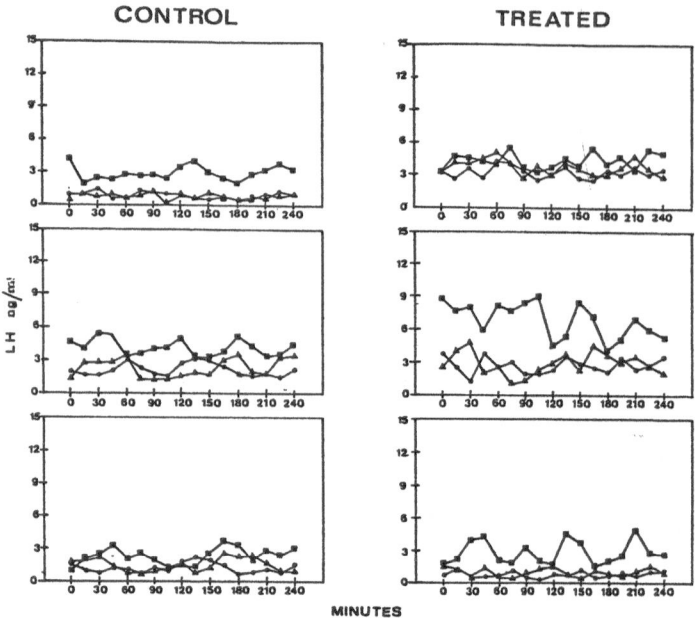

Fig. 3   Episodic release of LH in bromocriptine treated
and control sows 3 days prior to weaning (●——●), at weaning (▲——▲) and 3 days after weaning (■——■).

lactation could affect the sensitivity to GnRH of the pituitary gland of
lactating sows. The sows were injected with a single dose of GnRH (500 µg)
5 days after the start of bromocriptine treatment. Surprisingly the response to GnRH evaluated in terms of LH plasma levels, was less marked and
prompt in bromocriptine treated animals than in controls (Fig. 4). The
comparison of areas under the curve for LH of each animal by using Student's
unpaired T test, indicated that the bromocriptine treatment significantly
reduced (P < 0.05) the sensitivity of the hypophysis to GnRH stimulation.
We think that this inhibitory effect is exerted directly by the bromocripti_
ne and not by the drop of PRL since when PRL concentration was lowered by
removing the piglets no changes in LH plasma levels occurred (Bevers et al.,
1981). Moreover subcutaneous injection of the single high dose of bromocriptine produced a suppression of LH secretion in lactating sows (Kraeling et
al., 1982). The dopaminergic system has already been proved to inhibit the
secretion of LH in intact humans and ovariectomized rats and sheep (Weiner

et al.,1978; Hill et al.,1980; Owens et al.,1980). It is generally accepted
that this inhibition is mediated by a reduced secretion of GnRH, however
our findings suggest that the dopaminergic system can affect LH secretion
in the sow by acting at the pituitary gland level. These findings did not
help to explain the facilitating effect of bromocriptine treatment on the
resumption of the ovarian activity.

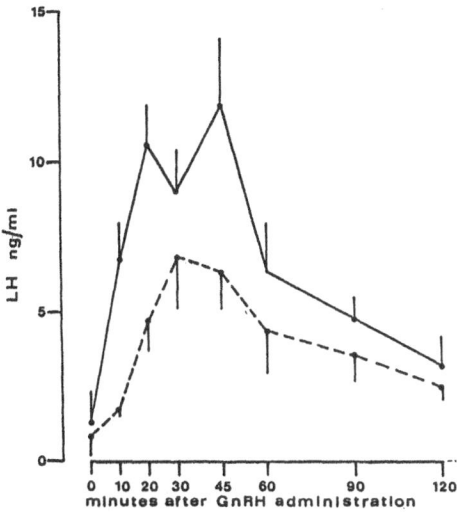

Fig. 4   Effect of bromocriptine treatment on GnRH stimulated
LH secretion. Four control (———) and four bromocriptine
treated sows (- - - -) were used. A single injection e.v. of
GnRH (500 μg) was administered 5 days after the start of bromocriptine treatment or at the corresponding time for control
sows.

Elsaesser and Parvizi (1980) showed that lactational anestrus in the
sows is characterized by a suppression of the estrogen positive feed-back
action on LH secretion. They found a partial recovery of this feed-back
mechanism only after 35 days of lactation. To evaluate if this inhibited
feed-back mechanism was the result of the high plasma levels occurring
during lactation we administered estradiol benzoate (EB) in sows 5 days
after the bromocriptine treatment had started (day 15 post-partum). Bromocriptine treatment and lactation continued until the fifth day after the
EB administration. This treatment produced a more marked increase of plasma
LH in bromocriptine treated sows than in control sows (Fig. 5). In parti-

cular 4 out of 4 treated sows presented a clear LH peak 50-80 hours after
EB injection while only 2 out of 4 control animals showed the same response.
However the higher sensibility of bromocriptine treated sows to estradiol
17β did not achieve statistical significance.

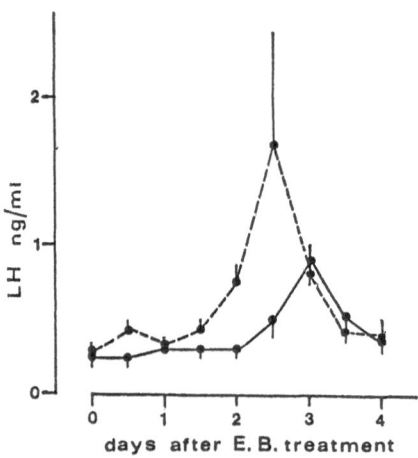

days after E. B. treatment

Fig. 5   Effect of estradiol benzoate administration on LH
plasma levels in 4 bromocriptine treated (▪ ▪ ▪ ▪) and 4 control sows (▬▬▬). EB was injected intramuscularly (50 μg/
Kg body weight) 5 days after the treatment had started or
at the corresponding period for control sows (20 days post-
partum). Lactation and treatment continued throughout the
experiment.

Therefore the facilitating effect of the bromocriptine treatment on
the resumption of ovarian activity might be ascribed to a more prompt
reactivation of the mechanisms which allow estrogens to stimulate LH secretion and thus initiate follicular development.

Following EB treatment a marked increase of PRL levels occurred in all
the bromocriptine treated animals, reaching values similar to those of
lactation (Fig. 6). This indicates that estradiol can reverse almost completely the inhibitory effect of a dopamine agonist like bromocriptine. A similar antidopaminergic effect of estradiol have been reported in the rat
(Labrie et al., 1980) while Williams and Ray (1980) could not find any positive effect of estrogens on PRL secretion in bromocriptine treated cows.

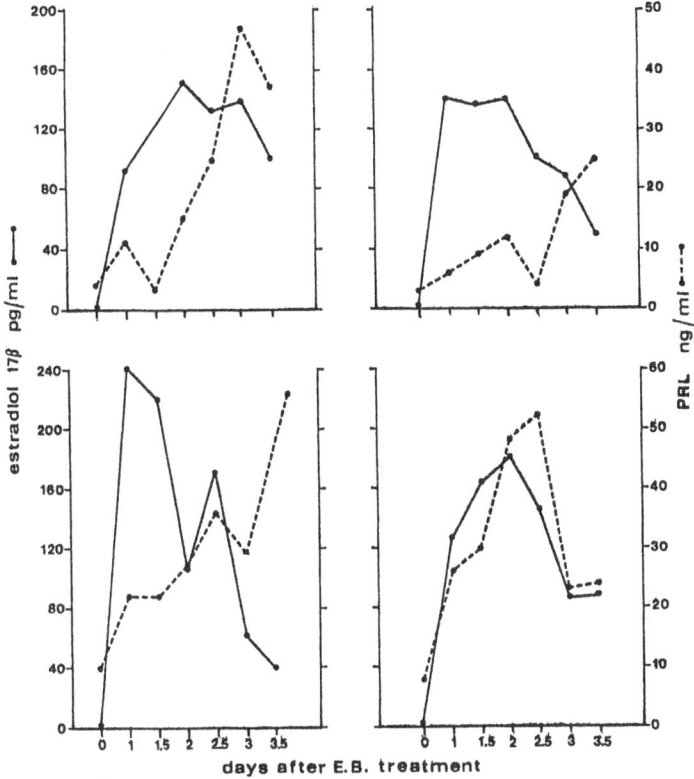

Fig. 6   Effect of EB on PRL secretion in 4 bromocriptine
treated sows.

In conclusion prolactin does not seem to be the principal factor responsible to the suppression of ovarian activity during lactation. However
the bromocriptine treatment could improve the reproductive efficiency of
post-partum sows. The increased stimulatory estrogen feed-back on LH secretion might account for this effect. However, in our experiments, the effect
of PRL withdrawal might have been masked by the bromocriptine treatment
which resulted capable of reducing the GnRH-stimulated LH secretion.

ACKNOWLEDGEMENTS

    We wish to thank Dr. M.M.Bevers for supplying anti pig PRL antiserum.
Prof.S.Chierichetti (Sandoz) for supplying bromocriptine and Prof. L.E.
Reichert for supplying standard preparations of pig LH and PRL. We are

grateful to the staff of Azienda Montroni for its assistance.

This work was supported by Regione Emilia Romagna and Ministero Pubblica Istruzione.

REFERENCES

Ash, R.W. and Heap, R.B. 1975. Estrogen, progesterone and corticosteroid concentrations in peripheral plasma of sows during pregnancy, parturition, lactation and after weaning. J.Endocr., 64, 141-154.

Bevers, M.M., Willemse, A.H. and Kruip, A.M. 1978. Plasma prolactin levels in the sow during lactation and the postweaning period as measured by radioimmunoassay. Biol. Reprod., 19, 628-634.

Bevers, M.M., Willemse, A.H., Kruip, A.M. and van de Wiel, D.F.M. 1981. Prolactin levels and the LH-response to synthetic LH-RH in the lactating sow. Anim.Reprod.Sci., 4, 155-163.

Bevers, M.M., Willemse, A.H. and Kruip, A.M. 1983. The effect of bromocriptine on luteinizing hormone levels in the lactating sows: evidence for a suppressive action by prolactin and the suckling stimulus. Acta Endocr., 104, 261-265.

Burger, J.F. 1952. Sex physiology of pig.Onderstepoort J.Vet.Res. Suppl.2, 3-218.

Corteel, J.M., Signoret, J.P., du Mesnil du Buisson, F. 1964. Variations seaisonnieres de la r production de la truie et facteurs favorisant l'anestrus temporale. ln Proceedings 5th Internat.Congr.Anim.Reprod. Artificial Insemination, 536-540.

Dsza, L. and Krzimowsk, H. 1981. Plasma prolactin levels in sows during pregnancy, parturition and early lactation. J.Reprod. Fert., 61, 131-134.

Edqvist, L.E., Einarsson, S. and Settergren, I. 1974. Ovarian activity and peripheral plasma levels of estrogens and progesterone in the lactating sows. Theriogenology, 1, 43-49.

Elsaesser, F. and Parvizi, N. 1980. Partial recovery of the stimulatory estrogen feedback action on LH release during late lactation in the pig. J.Reprod.Fert., 59, 63-67.

Hill, T.G., Alliston, C.W. and Malven, P.V. 1980. Plasma luteinizing hormone and prolactin in normothermic and hyperthermic ovariectomized ewes. Life Sci., 26, 1893.

Hurtgen, J.P., Leman, A.D. and Crabo, B. 1980. Seasonal influence on estrus activity in sows and gilts. JAVMA, 176, 119-123.

Kann, G., Martinet, J. and Schair, A. 1976. Inpairment of luteinizing hormone release following estrogen administration to hyperprolactinemic ewes. Nature, Lond., 264, 465-466.

Kann, G., Martinet, J. and Schair, A. 1977. Modifications of gonadotropin secretion during natural and artificial hyperprolactinemia in the ewe. In "Prolactin and human reproduction" (Academic Press, London). pp.47-59.

Kraeling, R.R., Rampacek, G.B., Cox, N.M. and Kiser, T.E. 1982. Prolactin and luteinizing hormone secretion after bromocriptine treatment in lactating sows and ovariectomized gilts. J.Anim.Sci., 54, 1212-1220.

Labrie, F., Ferland, L., Di Paolo, T. and Villeux, R. 1980. Modulation of PRL secretion by sex steroids and thyroid hormones. In "Central and peripheral regulation of prolactin function" (Ed.R.M.MacLeod and U. Scapagnini). (Raven Press,New York). pp.97-113.

Lauerdale, J.W., Kirkpatrick, R.L., First, N.L., Houser, E.R. and Casida L. E. 1965. Ovarian and pituitary gland changes in periparturient sows. J.Anim.Sci., 24, 1100-1103.

Love, R.J. 1978. Definition of a seasonal infertility problem in pigs. Vet. Rec., 103, 443-446.

Lu, K.H., Chen, H.T., Huang, H.H., Crandison, L., Marshall, S. and Meites, J. 1976. Relation between prolactin and gonadotropin secretion in postpartum lactating rats. J.Endocr., 68, 241-250.

Owens, R.E., Fleeger, J.C. and Harms, P.G. 1980. Evidence for central nervous system involvment in inhibition of luteinizing hormone release by dopamine receptor stimulation. Endocr. Res. Comm., 7, 99.

Parvizi, N., Elsaesser, F., Smidt, D. and Ellendorff, F. 1976. Plasma LH and progesterone in the adult female pig during the estrus cycle, late pregnancy and lactation and after ovariectomy and pentobarbitone treatment. J.Endocr., 69, 193-230.

Ravault, J.P., Martinat-Botte, F., Mauget, R., Martinat, M., Locatelli, A. and Bariteau, F. 1982. Influence of the duration of daylight on prolactin secretion in the pig:hourly rythm in ovariectomize female, monthly variations in domestic (male and female) wild strains during the year. Biol.Reprod., 27, 1084-1089.

Roland, R., Lequin, R.M., Schellerens, L.A. and de Jong, F.H. 1975. The role of prolactin in the restoration of ovarian function during the early post-partum in the human female. I. A study during physiological lactation. Clin. Endocr., 4, 15-25.

Smith, B.B. and Wagner, W.C. 1980. Lactation physiology in the pig. Proc. 9th Int.Congr.Anim.,Reprod.and A.I., Madrid, 3, 118.

Stevenson, J.S., Britt, J.H. 1980. Luteinizing hormone, total estrogens and progesterone secretion during lactation and after weaning in the sow. Theriogenology, 14, 453-462.

Stevenson, J.S., Cox, N.M. and Britt, J.H. 1981. Role of the ovary in controlling LH, FSH and prolactin secretion during and after lactation in pigs. Biol.Reprod., 24, 341-353.

Stevenson, J.S., Britt, H.J. 1981. Interval to estrus in sows and performances of pigs after alteration of litter size during late lactation. J.Anim.Sci., 53, 177-181.

Tarocco, C., Beccaro V. and Enne, G. 1979. Sterilità estiva dei suini. Suinicultura, 5, 65-72.

Taverne, M., Bevers, M., Bradshaw, J.M.C., Dielman, S.J., Willemse, A.H. and Porter, D.G. 1982. Plasma concentrations of prolactin, progesterone, relaxin and oestradiol 17ß in sows treated with progesterone, bromocriptine or indomethacin during late pregnancy. J.Reprod. Fert., 65, 85-96.

Thorner, M.O., Rogol, A.D., Evans, S.V., Nunley, V.C., McLeod, R.M. 1980. Effect of prolactin on gonadal function in man. In "Central and peripheral regulation of prolactin function" (Ed. R.M.MacLeod and U.Scapagnini). (Raven Press, New York). pp.271-286.

van Landeghem, A.J. and van de Wiel, D.F.M. 1978. Radioimmunoassay for porcine prolactin: plasma levels during lactation suckling and weaning and after TRH administration. Acta Endocr., 88, 653-657.

Weiner, R.I. and Ganong, W.F. 1978. Role of brain monoamines and histamine in regulation of anterior pituitary secretion. Physiol.Rev., 58, 905.

Williams, G.L. and Ray, D.E. 1980. Hormonal and reproductive profiles of early post-partum beef heifers after prolactin suppression on steroid induced luteal function. J.Anim.Sci., 50, 906-918.

Wuttke, W., Hohn, G.K., Honma, K., Hilgendorf, W. and Lamberts, R. 1980. Interrelationship between prolactin and gonadotropins. In "Central and peripheral regulation of prolactin function" (Ed. R.M.MacLeod and U. Scapagnini). (Raven Press, New York). pp.221-236.

DISCUSSION

Chairman: MAUGET, R. (France)

Beyond the seasonal effects on reproduction in the sow, lactational and subsequent postweaning anestrus are the causes of considerable loss of revenue in modern commercial pig husbandries. The underlying physiological mechanisms are not yet fully understood. It is intended that this session presents the current state of knowledge of the factors involved in the inhibition of ovarian activity during lactation.

The first presentation (G.R. Foxcroft/U.K.) assumes that hypothalamic-pituitary-ovarian changes observed after weaning reflect causes of lactational anestrus.Weaning removes the stimulus of lactation and reinstates follicular maturation. The endocrine status of the sow during lactation is not a stable one. As lactation advances, LH and FSH secretion progressively increases and escapes at weaning. This could result from a gradually decreasing blockage at the hypothalamic level.

The following paper (Ellendorff et al./F.R.G.) emphasized that suckling-related oxytocin secretion is a possible trigger for lactational anestrus. Suckling frequency and amplitude of oxytocin secretion progressively decrease with lactation. Evidence is presented to implicate oxytocin as a prevailing factor in the maintenance of anestrus during lactation.

The relative importance of prolactin in the inhibition of gonadotrophin secretion during lactational and postweaning estrus is discussed in the third paper (Booman et al./Netherlands). It seems that prolactin is only partially involved, acting with the suckling stimulus.

In a last paper, M. Mattioli and E. Seren (Italy) also investigate the role of prolactin by means of bromocriptine treatments. The withdrawal of prolactin does not significantly enhance the stimulating estrogen feedback.

Finally, our understanding of the physiological mechanism during lactation need further investigations. Since oxytocin appears to be one major contributing factor, it remains to determine its precise way(s) and site(s) of action. It is expected that further classification of mechanisms governing lactational anestrus in the sow will eventually provide biotechnical tools to markedly increase efficiency of pig production.

SESSION IV

POST PARTUM ANESTRUS

Chairman: H. Karg

POST PARTUM ANOESTRUS : PROBLEMS AND PERSPECTIVES

H. Karg

Lehrstuhl für Physiologie der Fortpflanzung und Laktation
Technische Universität München
8050 Freising-Weihenstephan
Federal Republic of Germany

Whether or not high lactation and the re-establishment of cyclicity are compatible characterizes differences between mammalian species and is discussed also for farm animals since many years (Smidt, 1966). We have learned during the preceding session of this meeting that in the sow - normally exhibiting exemplariously lactational anoestrus - minimizing the number of litters can result in oestrous cycles whilst maintaining lactation. Offsprings once per year can only be obtained in horses and cattle under the "regular" situation of simultaneous lactation and fertility. But it has been shown for both species that there is still competition in between these two physiological performances. For example lactation has an adverse effect on embryonic survival in horses (Merkt, 1966). The problems of the postpartum cow have been comprehensively reviewed during a former CEC-Seminar (Karg and Schallenberger, 1982). Three papers of today's session will bring the endocrinological aspects up to date.

But we should keep in mind that perspectives (necessarily extrapolated from retrospectives) should include interdisciplinary approaches examining the multifactorial problem of the postpartum situation. In about historical order we should glance over the aspects of (1) the breeders and animal husbandry economists, (2) the veterinary clinicians, (3) the nutritionists and (4) the endocrinologists.

ad 1) According to the experience that milk yield is in close relation to food supply which under conventional husbandry was depending on the season, breeding in large areas of Europe was frequently performed to obtain winter calvings. This was a benefit for the lactation curve and other reliable facts but as a consequence cyclicity had to resume in the months of the worst food supply (late winter and early spring) (Karg, 1966), which caused - if no supplementation according to modern nutrition physiology was provided - already increased demands for veterinary service.

ad 2) Veterinarians defined the biological rest-time and characterized uterine involution and re-establishment of ovarian function. Adverse effects of complications such as dystocia (Hoffmann et al., 1979), retained membranes and infections became a large field of activities (Vandeplassche and Bouters, 1982). We are very lucky to have with us the doyen of our clinicians, Prof. Vandeplassche, presenting lateron a comparative aspect on this topic.

ad 3) Nutritionists recognized a dramatic shift in balance occurring at parturition from the positive retention values of minerals, nitrogen, energy etc. ante partum to negative mobilization data post partum (Forbes, 1935; Lenkeit et al., 1962; Lenkeit et al., 1964; Oslage and Farries, 1966). Recommendations for feeding schedules ante versus post partum were elaborated. The effect of high or low energy supply before and after parturition on reproductive performances became obvious (Wiltbank et al., 1962; Wiltbank et al., 1964; Schilling et al., 1980). Multifactorial influences were encountered with the experience that the adverse effect of suckling or frequent milking on cyclicity can be counteracted by increasing levels of energy supply (Randel, 1978).

ad 4) The modern analytical repertoire of endocrinologists, which was established during the last 2 decades (Karg et al., 1976; Karg, 1983), allowed some insight into complicated mechanisms of regulation, especially concerning the pathways leading to the re-establishment of ovarian function post partum (Erb et al., 1971; Karg and Schallenberger, 1982; Peters and Lamming, 1984). It became obvious that the tremendous individualism in postpartum performance which is opposite to the precise sequences of other events of female reproduction like the oestrous cycle, pregnancy and parturition (Karg, 1972) has environmental but also endogenous causes. Recent investigations evaluating nonpathological factors exerting effects on the postpartum anoestrus period (Peters and Riley, 1982; Terqui et al., 1982; Pirchner et al., 1983; Karg and Schallenberger, 1983) pointed towards to following conditions: housing (outdoor movement versus tethering), season (shorter postpartum intervals at autumn and longer intervals at spring time), age (longer postpartum intervals at first calving, shorter at second calving), food supply, stress, breed, milk yield and maternal behaviour (longer interval if the own mother than foster cows get suckled). There is a certain correlation of milk yield and patho-physiological problems like silent heat and ovarian cysts (El-Keraby and Schilling, 1976).

Until now the geneticist remained almost unmentioned. But is there not a "true cow effect" ? Or with other words: has the progress in re-productive endocrinology not provided useful traits ? First steps using postpartum progesterone profiles for estimation of repeatabili-ties of postpartum acyclia and silent heat and for evaluation of biometrical relationships including the "true cow effect" have been undertaken (Pirchner et al., 1983).

Finally we should consider practical perspectives which could be summarized under 3 headings: 1. Management programs may take advantage of the obtained interdisciplinary informations and adapt it to the local requirements. This may concern adjustment of housing, nutrition schedules, supervision of herds for oestrus control, veterinary in-spections with prophylactic measures or treatments, monitoring success of artificial insemination by milk progesterone assay (Karg, 1981) etc. 2. Biotechnical measures may include if within the frame of economy also hormonal treatments to force resumption of cyclicity, twinnings etc. (O'Farrell, 1982; Roche et al., 1982; Sreenan, 1984). But it should be kept in mind that shortening the postpartum interval of a high yielding cow means also to shorten in part the benefit of an extended lactation curve (Esslemont, 1982; Kräusslich, 1982). 3. "Re-turn to nature". It may become more fashionable - and may even be helpful along the policy of the CEC to restrict the surplus of milk production - to create more areas for suckler and foster cow husbandry. Nursing a calf with cow's milk is then perhaps not any more deemed as a sophisticated approach.

REFERENCES

El-Keraby and E. Schilling. 1976. Feeding of cows during the late gestation on the reproductive performance in the early post partum period. Proc. VIIIth Intern.Congr. on Anim. Reprod. & A.I. July 1976, Cracow, Vol. III, 361-364.
Erb, R.E., Surve, A.H., Callahan, C.J., Randel, R.D. and Garverick, H.A. 1971. Reproductive steroids in the bovine. VII. Changes post partum. J. Anim. Sci. 33, 1060-1071.

184

Esslemont, R.J. 1982. Economic aspects related to cattle infertility and the postpartum interval. In: Factors influencing fertility in the postpartum cow. CEC-Seminar 1981. Eds. H.Karg and E.Schallenberger, M. Nijhoff Publishers, The Hague/Boston/London, 20 (Current topics in vet. med. and anim. sci.), 442-458.

Forbes, E.B. 1935. The mineral requirements of milk production. 319, Pennsylvania State College.

Hoffmann, B., Mason, I.L. and Schmidt J. 1979. Calving Problems and Early Viability of the Calf. CEC-Seminar 1977. 4 (Current topics in vet.med.and anim.sci.), M.Nijhoff Publ., The Hague/Boston/London.

Karg H. 1966. Fütterung und Fruchtbarkeit des weiblichen Rindes. Züchtungskd. 38, 400-412.

Karg H. 1972. Recent results concerning cyclic effects of pituitary gonadotropins. In Proceed. VIIth Intern.Congress on Animal Reprod. & Artif. Insem., Munich 1972, Vol I, 45-56.

Karg, H., Claus, R., Hoffmann, B.,Schallenberger, E. and Schams, D. 1976. Present status and future possibilities of radioimmunoassay in animal production. In: Nuclear Techniques in Animal Production and Health. Proceed. Intern.Sympos. Nuclear Techniques in Anim. Reprod. & Health, Wien, IAEA-SM 205/109, 487-511.

Karg, H. 1981. Physiological impact on fertility in cattle, with special emphasis on assessment of the reproductive function by progesterone assay. Livestock Prod. Sci. 8, 233-246.

Karg, H. and Schallenberger E. 1982. Factors influencing fertility in the postpartum cow. CEC-Seminar 1981. 20 (Current topics in vet. med. and anim. sci.), M. Nijhoff Publ., The Hague/Boston/London.

Karg, H. 1983. Applications of radioimmunoassays in veterinary medicine and animal science. In: Application of radioimmunoassay and related methods in animal science. Conf. Warschau, 1979; Ed. Panstwowe Wydawnictowo Naukowe, Polish Academy Nauk, Warszawa/Poland, 15-36.

Karg, H. and Schallenberger, E. 1983. Regulation der ovariellen Steroidhormonsekretion post partum. Wien. Tierärztl. Mschr. 70, 238-243.

Kräusslich, H., 1982. General conclusions and outlook of CEC-Seminar, Freising, 1981. In: Factors influencing fertility in the postpartum cow. Eds. H. Karg and E. Schallenberger, M. Nijhoff Publishers, The Hague/Boston/London, 20 (Current topics in vet. med. and anim. sci.), 574-578.

Lenkeit, W., Molnar, S., Lantzsch, H.J. and Gütte, J.O. 1962. Untersuchungen über den Beginn der Mobilisierung der eingelagerten Körperreserven von Ca, P und N im Graviditäts- und Laktationszyklus bei Sauen. Z.Tierphys., Tierern.u.Futtermittelkd.17, 65-90.

Lenkeit, W., Molnar, S., Risto, G. 1964. Veränderung der Harnzusammensetzung während des Graviditäts-Laktationszyklus bei Milchkühen. 2. Mitt.: Verlauf der C-Ausscheidungen und das C:N-Verhältnis im Harn gravider und laktierender Milchkühe. Z. Tierphysiol., Tierernährg. u. Futtermittelkd. 19, 291-297.

Merkt, H. 1966. Fohlenrosse und Fruchtresorption. Zuchthyg. 1, 102-108.

O'Farrell, K.J. 1982. Effects of management factors on the reproductive performance of the postpartum dairy cow. In: Factors influencing fertility in the postpartum cow. CEC-Seminar 1981. Eds. H. Karg and E. Schallenberger, M. Nijhoff Publishers, The Hague/Boston/ London, 20 (Current topics in vet. med. and anim. sci.), 510-529.

Oslage, H.J., Farries, F.E. 1966. Beiträge zum Stoffwechsel von Kühen im Ablauf von Trächtigkeit und Laktation. Landbauforschung Völkenrode 16, 53.

Peters, A.R. and Riley, G.M. 1982. Pulsatile LH secretion and its induction in post-partum beef cows. In: Factors influencing fertility
  in the postpartum cow. CEC-Seminar 1981. Eds. H. Karg and E.
  Schallenberger, M. Nijhoff Publishers, The Hague/Boston/London, 20
  (Current topics in vet.med. and anim. sci.), 225-228.

Peters, A.R. and Lamming, G.E. 1984. Endocrine changes in the postpartum
  period. In:Proceed. 10th Intern.Congr. Anim.Reprod. & A.I. June 1984,
  University of Illinois/Urbana-Champaign (USA), Vol.IV,
  III/17 - III/24.

Pirchner, F., Zwiauer, D., v. Butler, I., Claus, R. and Karg, H. 1983.
  Environmental and genetic influences on post-partum milk progesterone profiles of cows. Z. Tierzüchtg. u. Züchtungsbiol. 100, 304-
  315.

Randel, R.D. 1978. Special techniques for AI in Brahman cattle. In:
  Proceed. Annual Meeting Amer. Soc. for Theriogenology, 1978, Okla-
  homa City/ Oklahoma(USA), 113-117.

Roche, J.F., Edwards, S. and Niswender, G.D. 1982. Induction of ovulation in post-partum beef cows. In: Factors influencing fertility
  in the post-partum cow. CEC-Seminar 1981. Eds. H. Karg and
  E.Schallenberger, M. Nijhoff Publishers, The Hague/Boston/London,
  20 (Current topics in vet.med. and anim.sci.), 530-535.

Schilling, E., Smidt, D., Farries, E. and Gauchel, F.R. 1980. Different
  prepartum feeding levels in dairy cows and the post partum reproductive efficiency. Proceed. 9th Intern.Congr. Anim.Prod. & A.I.,
  Madrid 1980, Vol. V, 283-286.

Smidt, D. 1966. Funktionale Beziehungen zwischen Fortpflanzungsorganen
  und Milchdrüse. Züchtungskd. 38, 271-282.

Sreenan, J.M. 1984. Steroid immunization in cows: Potential for increa-
  sing ovulation and twinning rates. In: Proceed. 10th Intern.
  Congr. Anim. Reprod. & A.I., June 1984, University of Illinois/
  Urbana-Champaign (USA), Vol. IV, VIII/22 - VIII/27.

Terqui, M., Chupin, D., Gauthier, D., Perez, N., Pelot, J. and Mauleon,
  P., 1982. Influence of management and nutrition on postpartum
  endocrine function and ovarian activity in cows. In: Factors
  influencing fertility in the postpartum cow. CEC-Seminar 1981.
  Eds. H.Karg and E.Schallenberger, M. Nijhoff Publishers, The
  Hague/Boston/London, 20 (Current topics in vet. med. and
  anim.sci.), 384-408.

Vandeplassche, M. and Bouters, R. 1982. The impact of gynaecological
  and obstetrical problems resulting out of pregnancy and parturition. In: Factors influencing fertility in the postpartum cow.
  CEC-Seminar 1981. Eds. H. Karg and E. Schallenberger, M. Nijhoff
  Publishers, The Hague/Boston/London, 20 (Current topics in vet.
  med. and anim.sci.), 30-44.

Wiltbank, J.N., Rowden, W.W., Ingalls, J.E., Gregory, K.E., Koch, R.M.
  1962. Effect of energy level on reproductive phenomenon of mature
  hereford cows. J. Anim. Sci. 21, 219-225.

Wiltbank, J.N., Rowden, W.W., Ingalls, J.E. and Zimmermann, D.R. 1964.
  Influence of post-partum energy level on reproductive performance
  of hereford cows restricted in energy intake prior to calving. J.
  Anim. Sci. 23, 1049-1053.

COMPARATIVE ASPECTS OF THE POSTPARTUM PERIOD IN DOMESTIC ANIMALS

M. Vandeplassche
Department of Reproduction and Obstetrics
Faculty of Veterinary Medicine
State University, Gent, Belgium

ABSTRACT

Investigations and observations in clinical patients are confronted with data from literature in order to find the indicated methods for evaluating the exact postpartum uterine involution and ovarian activity. A comparative description of the characteristic features of the postpartum period is given in the mare, cow, ewe, sow, bitch and rabbit. An analysis is made of the influence of 11 different factors which can shorten or prolong the length of the postpartum anoestrous period and parturition interval comparatively in the 6 animal species.

Those factors are : the climate and season, lactation or suckling period and early weaning, animal breed, body condition and energy intake, diseases, individual genetic variation, age and body weight, sexual interplay, stress, physical training, and management. The reactions to those influencing factors vary markedly between different animal species and could be helpful to prevent unjustified generalizations ; they also may contribute to open the way for important practical applications.

INTRODUCTION

The postpartum period in domestic animals is characterized by a quite variable period of ovarian subactivity or inactivity or anoestrus resulting in a variable length of the parturition interval. The latter has important economic repercussions in the breeding of those farm animals in which an efficient animal production needs an intense reproduction.

Looking for an exact definition of the postpartum period, one could consider the period from parturition to complete genital tract involution, or to the first oestrus, or to the first ovulation(s), or to the first natural mating or even to the reconception.

The aim of this paper will be to analyse and to compare the effect of the most important factors influencing the course of the postpartum period in the mare, the cow, the ewe, the sow, the bitch and the rabbit. Such a comparative study may help to prevent generalization for all animal species from observations in one or two species and it could also give an opportunity for comparative study of special aspects for a better understanding of some biological mechanisms.

MATERIALS AND METHODS

Data from personal experience with a great number of clinical patients and of experimental animals (mainly mares, cows and sows),have been compared with and completed by data from the literature (more particularly concerning the ewe, the bitch and the rabbit).

Several methods of examination have been used in evaluating the postpartum involution and ovarian activity :
1. Detection of heat (oestrous behaviour)
2. Rectal palpation in large species of the cervix, uterus, ovaries (presence of follicles, ovulation or corpus luteum), vascular fremitus
3. Vaginal examination can be performed clinically for the observation of relaxation, hyperaemia and mucus, whereafter the latter can be studied cytologically in the bitch and sow
4. Hormonal analysis in blood or milk of progesterone and oestrogens
5. Laparotomy and postmortem examination of uterus and ovaries.

OBSERVATIONS AND RESULTS

A. Characteristics of the postpartum involution and ovarian activity in different domestic animal species.

Taking in consideration the marked species differences in respect of the influence of suckling, the postpartum period will be compared here in non-suckled animals.

The mare
- rapid uterine involution during 10-14 days pp (Vandeplassche, 1981 ; Vandeplassche et al., 1983). This is improved by obstetric asepsis and minimal trauma, by uterine contraction and by maximal microbial phagocytosis (Vandeplassche, 1984).
- the first pp oestrus at about day 9 to 12 is ovulatory with overt heat. This is hardly influenced by the presence of the foal and by suckling.

- the curves of LH, oestrogens and progesterone and luteal lifespan are similar to those in non-parturient cycling mares (Ginther, 1979), and there is almost normal fertility at foal heat (Spincemaille et al., 1980).

- when regularly teased with a stallion, oestrous receptivity is often present about 10 days after foal heat. This could be caused by a mid-cycle follicle (Vandeplassche et al., 1978) or by a shortened luteal lifespan.

The cow
- slow uterine involution during 4 weeks pp (Vandeplassche 1981). This is delayed by trauma, infection and damage to the phagocytosis (Vandeplassche, 1984).

- the first ovulation happens between day 15 and 20 pp

- 50% of the first pp corpora lutea are deficient, causing low progesterone bloodlevel, shortened cycle and silent heat.

- the second pp ovulation occurs near day 35. The resulting corpus luteum and cycle are almost normal (King et al., 1976 ; Troxel-Kesler, 1984).

The ewe
- slow uterine involution during 3-4 weeks pp
- during the anoestrous pp period, there is some follicular growth but rarely ovulation and no overt heat
- the first pp ovulation mainly depends on season and breed. Often there is silent heat (Lawson et al., 1984).

The sow
- fairly rapid uterine involution during 2-3 weeks. This is improved by uterine contraction and intense phagocytosis (Vandeplassche, 1981).

- often oestrous behaviour is shown at 3-4 days pp, but no ovulation occurs (Burger, 1952).

- in the pp period some follicular development happens (Kudlac and Groch, 1979) and small surges of oestrogens are present in some sows (Edqvist et al., 1973).

The bitch

- very slow uterine involution during 4 months pp
- long oestrus : the first ovulatory cycle is at 4 months pp (Shille and Stabenfeldt, 1980 ; Andersen and Simpson,1973).

The rabbit

- rapid uterine involution pp
- immediately pp there are mature follicles and the animal is receptive for mating. The latter induces normal ovulations, luteal lifespan and fertility. There is little influence of suckling (Foxcroft and Hasnain, 1973 ; Torres et al., 1977).

B. General remarks concerning the postpartum period in different domestic animals.

1. The corpora lutea of pregnancy either disappear in the course of gestation (mare) or they become inactive shortly before parturition and regress (cow, ewe, sow, bitch, rabbit).

2. Examples of normal and of prolonged postpartum anoestrus and parturition interval in cattle.

Table 1 Calving interval in dairy cattle with A.I. in a temperate climate (The Netherlands, Annual Report A.I., 1983)

- average for the whole country	380 d.
- for excellent herds	359 d.
- for problem herds	439 d.

Table 2 Calving interval in dairy cattle with natural breeding in a subtropical climate : Sri Lanka (Vandeplassche and Kumaraswamy, 1975 ; report F.A.O. Rome)

- for excellent herds (10%)	: 12-13 months
- for two problem herds	: 14.5 months for primiparous and multiparous cows
	16.4 months for primiparous alone

Data of 2 herds, 140 cows, period of 2 years.

Table 3   Calving interval in herds of dairy cattle in a tempe-
          rate climate (Belgium, De Backer, M., in "De Boer",
          n°36, 4 Sept. 1984) : 100 herds over 5 years

Length of interval in months (m)	percentage (%) of herds
15 m.	8.2%
14 m.	21.4%
13 m.	39.9%
12 m.	30.5%

Those examples may give an idea of the surprisingly high
incidence and economic importance of prolonged postpartum anestrus and calving interval in dairy cattle, in which ideal
production is positively correlated with calving intervals
shorter than 13 months (De Kruif, 1975).

C. Factors influencing the postpartum period in different
domestic animals.

The pp anoestrous period may vary under the influence of
several factors which will be discussed in order of decreasing
importance : season, lactation, breed, nutrition, diseases, individual sensitivity, age, sexual contact, stress, physical
training, management.

However, several of those factors may interact one with
another, thus complicating a sharp circumscription (e.g. season - nutrition diseases - management ; breeds - diseases -
stress or lactation ; nutrition - diseases).

Nevertheless a trial will be made to compare the effect of
11 different factors on the pp period in each of the 6 animal
species mentioned above.

1. The climate and season

- This is a complex of mainly length of daylight (photoperiodic alterations), temperature and some interacting factors in temperate and tropical countries.

- The domestic animal species can be classified in
order of decreasing sensitivity to seasonal ovarian activity :
mare, ewe > rabbit > sow (Claus et al., 1984)> cow (Lamming

et al., 1981 ; Arthur et al., 1982 ; Hansen and Hauser, 1984)
> bitch.

2. The lactation
- This is a complex of different factors with inhibitory effect on the pp ovarian activity. Classification of these factors in order of decreasing importance : presence of own offspring (Terqui et al., 1982) > presence of foreign kids suckling > nutritional stress > milking (Wagner and Li,1982).
- The inhibitory effect of the lactation on the pp length of anoestrus varies in the different species. This can also be classified in decreasing order : sow> ruminants (Casida et al., 1968) > bitch > rabbit, mare.
- One suckling lamb or calf inhibits less than 2 or 3 do.
- (Very) early weaning of piglets markedly influences the subsequent fertility of the sows. Van der Heyde studied over 10 years the effect of weaning at 2 weeks pp in a herd of 120 sows. These sows were selected and cross-bred for fertility, and copiously fed during the gestation, lactation and further pp period. He described the following observations :
- Litter size increased from 7.8 to 11.5 pigs/litter and 20-25 piglets/year/sow
- First breeding at weight of $\pm$ 125 kg and 2nd or 3rd oestrus
- Piglet mortality from weaning until 25 kg : 0.8 to 1.5%
- Spontaneous postweaning oestrus : 80% within 14 days pp
- Pregnancy rate : 80%
- Maternal longevity : 5.7 litters/sow
- The comparative effect of weaning, milking and suckling in beef cattle of the breed of Middle and High Belgium is described by Derivaux (1984) : the postpartum anoestrous period lasted 23 days in 5 cows weaned immediately postpartum, 45 days in 20 cows milked twice a day, and 115 days in 5 cows freely suckled by one calf.

3. The animal breed
- purebred : mares differ from ponys, double muscled cows from beef cattle (Vandeplassche, 1974) and beef cattle from dairy cows, while local breeds differ from imported cows and sheep.

- crossbred : cows and sows have a shorter parturition interval than purebreds (King and MacLeod, 1984) and than inbred breeds.

4. The body condition and energy intake.

- in the second half of gestation and pp, high level feeding and flushing results in early pp oestrus and shortens parturition interval in mares (Henneke et al., 1984), cows, sows (Wrathall, 1975) and ewes (King, 1978). This effect is most obvious in case of poor body condition, e.g. after twin gestation in cows.

- a certain minimum body weight is more important than the loss of weight (Table 4).

- underfeeding affects more primiparous than multiparous cows (Vandeplassche and Kumaraswamy, 1975 ; Tervit et al., 1977) and sows.

Table 4   Correlation of reconception to body weight in beef cattle in a tropical country (Grosskopf, 1978,South-Africa) : data of 2134 cows & heifers over 12 years.

Body weight 1 day postpartum	weight at end of breeding season	loss of weight	conception rate
530 kg	470 kg	60 kg	64%
480 kg	420 kg	60 kg	36%

5. The diseases

- the pp period is influenced by every severe general disease such as fever, skin diseases or blood parasites (Bodhipaksha, 1984).

- pp anoestrus is prolonged by diseases of the genital tract such as dystocia and endometritis (De Sutter, 1954 ; Morrow et al., 1969 ; Vandeplassche, 1973 ; Mather and Melancon, 1981) and cystic ovaries (Vandeplassche et al.,1971 ; Bostedt, 1979)

- a herd health program is therefore very important.

6. The individual animal

- the genetic base of each animal results in a variable predisposition for or resistance against unfavourable environmental factors such as climate, nutrition, age, stress etc. (King et al., 1976).

7. The age and body weight

- at first weaning, there is a significant correlation between these factors and the time-lapse between the weaning and the first pp oestrus in sows (Lievens-Van der Heyde, 1984)

- primiparous cows have the first pp oestrus 1-2 weeks later than multiparous cows (Tervit et al., 1977 ; Lamming et al., 1981).

8. Sexual interplay

- sexual interplay shortens the pp anoestrus in the cow, sow and ewe

- this phenomenon is shown in females by mutual mounting and standing (King et al.,1976), and in contact of females with vigorous males (Whitten,1956) ; it is provoked by way of pheromones (sex attractants) from glands, wool, urine, faeces, and tactile stimuli from body contact (Knight et al.,1983)

- the presence of several boars is more efficient than that of one (Brooks and Cole, 1970)

9. The stress

- when stress is intense and frequent, pp anoestrus is longer in sensitive animals

- mild stress of nursing postpones the first pp anoestrus in the cow and sow (Wagner and Li, 1982)

10. The physical training

- loose housing and forced training improve uterine involution and shorten pp anoestrus in cows (Schipilow,1965 Bostedt et al., 1984).

11. Management

- this is a complex combination of multiple factors involving both the herdsman and the herd

- the professional skill of the herdsman largely influences the proper feeding, housing, breeding techniques, heat detection and disease control

- larger herds are more difficult to supervise than small herds (Lamming et al., 1981).

DISCUSSION AND CONCLUSIONS

For a reliable diagnosis of the pp ovarian activity, clinical methods of examination (including for cows the palpation of

the uterine artery for judging its size and eventual presence
of fremitus)have to be completed by cytological and hormonal
examinations, eventually by laparotomy and necropsy.

The fact should be stressed that inactivation and regression of the corpora lutea of pregnancy start shortly ante partum in all domestic animals.

The uterine involution is very rapid in rabbits and mares,
fairly fast in sows, slow in cows and ewes and extremely slow
in the bitch. The first pp ovulatory oestrus is immediate in
rabbits, is quite fast in the mare, it is slower in sows and
especially in cows where it may result in an economically important prolonged calving interval in tropical (Table 2) and
in temperate climates (Table 1, 3).

Despite some interactions, 11 different factors can influence the pp anoestrous period. The climate and season have a
deciding influence in mares and ewes, but little in cattle.Nur-
sing inhibits completely pp ovulation in sows (importance of
early weaning in sows and cows), and very little in the mare
and rabbit. Crossbreeding shortens pp anoestrus compared with
purebreeding and inbreeding. In respect of nutrition and energy intake, a certain minimum of body condition is more important than the loss of weight (Table 4), especially for primiparous cows and sows. Diseases can strenghten the effect of
underfeeding and may also directly postpone the first pp ovulation as it happens in about 25% of the cows and over 90% of the
sows suffering from cystic degeneration of the ovaries and also
in cows with pp endometritis. The latter is often a consequence
of trauma and vascular thrombosis, retained placenta, delayed
uterine involution, increased infection and poor bacterial phagocytosis. This phagocytosis is more efficient in the mare and
sow than in the cow (Vandeplassche and Bouters, 1983). It is
severely damaged by butazolidines and by antiseptics, and can
be stimulated by oestrogens, because the blood-progesterone-
level is often positive (Vandeplassche and Coryn, 1980),and by
injection of incomplete Freund's adjuvans (Vandeplassche,1984).
An alternative treatment of severe pp endometritis in mares and
cows is the intra-uterine infusion of 150 ml heparinized blood
of the animal : the phagocytizing capacity of blood neutrophils

is about ten times stronger than that of neutrophils from uterine exudate (Vandeplassche and Bouters, 1983). This factor is
again supported by the blood-complement and by specific antibodies against streptococci and corynebacteria (Schulz et al.,
1979). The quality, but also the quantity of sexual interplay
are quite efficient stimuli in shortening the pp anoestrus in
most species. Housing and management should take that in consideration. Free running and physical training in the pp period greatly prevent severe stress and support the general
health.

ACKNOWLEDGMENT
    The author is most grateful to Dr. P. Simoens for his va-
luable remarks in the preparation of this manuscript.

REFERENCES

Andersen, A.C. and Simpson, M.E. 1978. The ovary and reproduc-
        tive cycle of the dog (Beagle). Geron-X Inc. Los Altos,
        California, 290 pp.
Arthur, G.H., Rahim, A.I.A. and Ismail, O.E. 1982. Fruchtbar-
        keitsüberwachung von europäischen Importzuchttieren in
        Saudi Arabia. Fachtagung Wels, Austria.
Bodhipaksha, P. 1984. Clinical cases of reproductive disorders
        in livestock in Thailand. Proc. Tropentag, Mai 1984, Univ.
        Giessen (D).
Bostedt, H. 1979.
        Berl. Münch. tierärztl.Wochschr., 92, 43-47.
Bostedt, H. Kozicki, L.E., Arnstadt, K.J. und Finger, K.H.1983.
        Einflusz von Haltungsbedingungen bei Milchkühen. Zuchthyg.
        18, 123-124.
Brooks, P.H. and Cole, D.J.A. 1970. The effect of the presence
        of a boar on the attainment of puberty in gilts. J.Reprod.
        Fert., 23, 435-440
Burger, J.F. 1952. Sex physiology of pigs. Onderstepoort J.
        Vet. Res., Suppl.2, pp 218.
Casida, L.E., Graves, W.E., Hauser, E.R., Lauderdale, J.W.,
        Reisen, J.W., Saiduddin, S. and Tyler, W.J. 1968.Studies
        on the postpartum cow. Res. Bull. Agric.Exp. Stn., Univ.
        Wisconsin n°270, 54 pp.
Claus, R., Schelkle, G. und Weiler, U. 1984. Erste Versuche zur
        Verbesserung der Fruchtbarkeitslage von Sauen im Sommer
        durch ein Lichtprogramm. Zuchthyg., 19, 49-56.
De Backer, M. 1984. Hoge producties en vruchtbaarheid:rundvee.
        De Boer, nr.36, p.25 (B).
De Kruif, A. 1975. Fertiliteit en subfertiliteit bij het vrou-
        welijk rund. Thesis, Utrecht.
Derivaux, J., Beckers, J. et Ectors, F.1984. L'anoestrus du
        postpartum. Vl.Diergeneesk.Tijdschr., 53, 215-229.

De Sutter, E. 1954. Puerperale stoornissen na keizersnede bij runderen. Vl.Diergeneesk.Tijdschr., 23, 273-286.

Edqvist, L.E., Einarsson, S. and Settergren, I. 1974. Ovarian activity and peripheral plasmalevels of oestrogens and progesterone in the lactating sow. Theriogenology, 1, 2.

Foxcroft, G.R. and Hasnain, H. 1973. Effect of suckling and time to mating after parturition on reproduction in the domestic rabbit. J.Reprod.Fert., 33, 367-377.

Ginther, O.J. 1979. Reproductive biology of the mare. Mc Naughton & Gunn Inc. Ann Arbor, Michigan, pp. 413.

Grosskopf, J.F.W. 1978. Non pathogenic factors associated with the reconception of beef cows under extensive conditions. IV. World Conf. Anim. Prod., Buenos Aires.

Hansen, P.J. and Hauser, E.R. 1984. Photoperiodic alteration of postpartum reproductive function in suckled cows. Theriogenology, 22, 1-14.

Henneke, D.R., Potter, G.D. and Kreider, J.L. 1984. Body condition during pregnancy and reproductive efficiency in mares. Theriogenology, 21, 897-909.

Henry, M.R., 1981. Some special aspects of the physiopathology of reproduction in mares. Thesis, Gent(B), pp. 219.

King, G.J., Hurnik, J.F. and Robertson, H.A. 1976. Estrous behaviour and ovarian function in dairy cows under intensive systems of management. Proc. 8th Int. Congr.Anim.Reprod.& A.I. Krakow, III, 149-152.

King, G.K. and McLeod, G.K. 1984. Reproductive function in beef cows calving in the spring or fall. Anim.Reprod.Sic.,6, 255-266.

Knight, T.W., Tervit, H.R. and Lynch, P.R. 1983. Effects of boar pheromones, ram's wool and presence of bucks on ovarian activity in anovular ewes early in the breeding season. Anim.Reprod.Sci., 6, 129-134.

Kudlac, E. and Groch, L. 1979. Morphologische Veränderung an Ovarien und Uterus bei Sauen nach Besamung. Zuchthyg.,14, 64-72.

Lamming, G.E., Wathes, D.C. and Peters, A.R.1981. Endocrine patterns of the postpartum cow. J.Reprod.Fert., suppl.30, 155-170.

Lawson, J.L., Forrest, D.W. and Shelton, M. 1984. Reproductive response to suckling manipulation in spanish goats. Theriogenology, 21, 747-755.

Lievens, R. and Van der Heyde, H. 1984. Reproductive performance according to weight and age of the sow at the end of the first lactation. in : Results of pig research - IWONL, De Crayerstraat 6, Brussels, pp. 291.

Mather, E. and Melancon, J.J. 1981. The periparturient cow. A pivotal entity in dairy production. J. Dairy Sci., 64, 1422-1430.

Morrow, D.A., Roberts, S.J. and McEntee, K. 1969. Postpartum ovarian activity and involution of the uterus and cervix in dairy cattle. I. Ovarian activity. Corn.Vet., LIX, 173-190.

Peters, A.R. and Lamming, E. 1984. Endocrine changes in the postpartum period. Proc.10th Int.Congr.Anim.Reprod. & A.I. Urbana, Illinois, USA, vol. IV, III 17-24.

Schäkel, W. und Ellendorff, F. 1984. Oxytocin verlängert den

Laktationsanöstrus beim Schwein. Zuchthyg., 19, 113.

Schipilow, S.W. 1963. Neue Tatsachen über die Erhöhung der Fruchtbarkeit und Leistung der Kühe. Zuchthyg., 7, 1-3.

Schulz, J., Van Aert, R., Dekeyser, P. und Vandeplassche, M. 1979. Immunofluoreszenzuntersuchungen zum Nachweis von Antikörpern gegen Corynebacterium pyogenes und Streptokokken im Blutserum und Vaginalschleim von Rindern. Arch. Exp. Vet. Med., Leipzig, 33, 783-789.

Schille, V.M., Stabenfeldt, G.H. 1980. Current concepts in reproduction of the dog and cat. Advances in Vet. Sci. & Comp. Med., 24, 211-244.

Spincemaille, J., Vandeplassche, M. and Tijskens, R. 1980. The comparative fertility in mares served or not at foal-heat. Proc. 9th Int.Congr. Anim. Reprod. & A.I., Madrid, IV, 221-224.

Terqui, M., Chupin, D., Gauthier, D., Perez, N., Pelot, J. and Mauleon, P. 1982. Influence of management and nutrition on postpartum endocrine function and ovarian activity in cows. in "Current Top. Vet. Med. Anim. Sci.", 20, 384-408. C.E.C.

Tervit, H.R., Smith, J.F. and Kaltenbach, C.C. 1977. Postpartum anoestrus in beef cattle : a review. Proc. N.Z. Soc. Anim. Prod. 37, 109-119.

Troxel, T.R. and Kesler, D.J. 1984. The effect of progestin and GnRH treatments on ovarian function and reproductive hormone secretions of anestrous postpartum suckled beef cows. Theriogenology, 21, 699-709.

Torres, S., Gerard, M., Thibault, C. 1977. Fertility factors in lactating rabbits mated 24 hours and 25 days after parturition. Ann.Biol.Anim.Bioch.Phys., 17, 63-69.

Vandeplassche, M., Spincemaille, J. und Bouters, R. 1971. Die zystöse Eierstocksdegeneration bei der Sau. Deutsch.tierärztl. Wochschr., 78, 91-93.

Vandeplassche, M. 1974. Die Vererbung des Merkmals "Doppelender" und dessen Bedeutung für die Rinderfleischproduktion. Der Tierzüchter, 26, 335-338.

Vandeplassche, M. and Kumaraswamy, S. 1975. The calving interval in heifers and in cows in Sri Lanka. in Report F.A.O. Rome.

Vandeplassche, M., Henry, M. and Coryn, M. 1978. The mature midcycle follicle in the mare. J.Reprod.Fert.,Suppl.27, 1-6.

Vandeplassche, M. and Coryn, M. 1980. Der Blutprogesteronspiegel im frühen Puerperium von Kühen. Monatsh.Vet.Med., 35, 425-429.

Vandeplassche, M. 1981. Neue vergleichende Aspekte der Involution und der puerperalen Metritis bei Stute, Kuh und Sau. Monatsh. Vet. Med., 36, 804-807.

Vandeplassche, M. and Bouters, R. 1983. Phagocytosis in the blood and uterine exudate of mares, cows and sows. Proc. 3rd Int.Symp.World Ass. Vet. Lab. Diagn., 1, 83-89.

Vandeplassche, M., Bouters, R , Spincemaille, J., Bonte,P. and Coryn, M. 1983. Observations on involution and puerperal endometritis in mares. Irish Vet. J., 37, 126-132.

Vandeplassche, M. 1984. Stimulation and inhibition of phagocytosis in domestic animals. Proc. 10th Int.Congr. Anim. Reprod. & A.I., Urbana, Illinois,USA, III, brief comm.475.

Wagner, W.C. and Li, P.S. 1982. Influence of adrenal corticoste-
    roids on postpartum pituitary and ovarian function. In
    "Current Top. Vet. Med. Anim. Sci.", 20, 197-219.
Whitten, W.K. 1956. Modification of mouse oestrous cycle by
    external stimuli associated with the male. J. Endocrin.,
    13, 399-404.
Wrathall, A.E. 1975. Reproductive disorders in pigs. C.A.B.,
    Ministry of Agric., Weybridge, Engl. pp. 313.

REPRODUCTIVE POTENTIAL DURING THE POST PARTUM PERIOD IN COWS

M. Terqui

Institut National de la Recherche Agronomique
Station de Physiologie de la Reproduction
Nouzilly, 37380 Monnaie, France

ABSTRACT
    The resumption of ovarian activity is delayed by various factors and most of the nursing females are in ovarian inactivity when they have to be bred. Teatments such as the introduction bull, temporary calf removal, progestagen and PMSG gave variable results of induction of oestrus and ovulation. Endocrine studies suggest that there are two main steps in ovarian inactivity weak (FSH+, LH-) and deep(FSH-, LH-). The variable percentage of females in deep or in weak inactivity presumably explains the variability of the response to treatments.

INTRODUCTION

    A one year interval between calving is the goal of the reproductive managment in beef and dairy herds, this is difficult to reach and to maintain because the restoration of a full ovarian activity (oestrus and fertile ovulation) is delayed by various factors. Oestrus and ovulation have to be induced. However, the treatments of induction are not efficient in all females in ovarian inactivity. This suggests that ovarian inactivity is not a steady state (ROCHE and IRELAND, 1984; TERQUI and LEGAULT, 1984).

    Thus, the objective of this paper is to examine the ovarian activity and the factors which control it in bovine, the different treatments elaborated to induce ovulation and an endocrine definition of ovarian inactivity.

I.- OVARIAN ACTIVITY

    Resumption of ovarian activity after calving has been extensively studied, mainly in dairy cows, with progesterone levels in plasma or in milk. As shown in Table I, five main groups have been recognized in France (TERQUI et al., 1982; PELOT et al., 1984).

During breeding period, all dairy heifers were cyclic but only 80p. cent of dairy cows exhibited ovarian activity 45 to 60 days post calving.

TABLE I   Mean ovarian activity of the main groups of bovine females, in France, during their normal breeding period.

Groups	Ovarian activity in p.100	Breeding period
Dairy heifers	95-100	Nov. to Dec.
Dairy cows	80	Dec. to May
Beef and rustic heifers	50	Feb. to May
Rustic cows	25	Feb. to May
Beef cows	15	Feb. to May

One out of two rustic or beef heifers was cyclic when the farmer wishes to breed them. The group of rustic cows like Salers and Aubrac has a poor level of ovarian activity (25p. cent) fourty to sixty days after calving. Nevertheless, the group with the lowest ovarian activity is the group of beef cows such as Charolais cows.

The values reported in Table I are means and the percentage of females in ovarian activity was very variable from farm to farm. Many factors have been identified as acting on cyclic ovarian activity: breed, sire of the foetus of the previous pregnancy, suckling, maternal behaviour parity, season, calving date, undernutrition, dark surrounding, tethering (TERQUI et al., 1982; THATCHER et al., 1982; PELOT et al., 1984; THIMONIER and GAUTHIER, 1984).

Among these factors the more important is probably underfeeding since most of beef and rustic females experiences periods of undernutrition. There is a graduation in the effect of undernutrition according to the body condition after calving, for example. Managment has also large consequences on the resumption of ovarian activity. When Salers cows were tethered in dark surrounding, the dynamics of the resumption of ovarian activity of such herd was much lower and slow than that observed for loose housed herd with daylight conditions (PELOT et al., 1984).

Thus, genetic, physiological, environmental and managment factors create a wide range of situations in herd ovarian activity level.

II.- INDUCTION OF OESTRUS AND OVULATION

Two different approaches have been taken to induce oestrus and ovulation, changes in management and hormones administration.

* The introduction of bull with females after calving reduced significantly both the interval between calving and first ovulation and its variance in the group of females with normal levels of nutrition (Table 2).

TABLE II The effect of bull and nutrition levels on the interval between calving and first ovulation (from MONJE et al., 1983).

| | Interval in days | |
	Normal level	Low level
Without bull	78	> 110
With bull	58	> 110

But when females were restricted the interval was increased and the introduction of the bull was not efficient to induce ovulation in restricted females.

Temporary weaning of calf allows also a reduction of the number of days open, but again this method of induction did not seem to be efficient when nutrition level was low (ALBERIO, 1985, Pers. Comm.).

* Hormonal treatments for induction and synchronization of oestrus and ovulation involve a progestagen and a PMSG administration (PELOT et al., 1984; ROCHE and IRELAND, 1984). When Norgestomet implant is inserted, 3 mg of Norgestomet and 5 mg of oestradiol 17 -valerate in oil solution were injected, the implant is left for 9 or 10 days. At time of implant removal, PMSG is injected. The dosage of PMSG is adjusted to female groups as indicated in Table 3. In dairy cows, PGF_2 analogue is injected two days before implant removal. The female were inseminated once at fixed time 48 or 56 hours after implant removal for heifers and cows respectively (PELOT et al., 1984).

Table III Ovulation and fertility after Norgestomet treatment

Group of females	Dose of PMSG in I.U.	P.cent of females without ovulation after treatment	Fertility as calving rate in p.cent
Dairy heifers	0	-	60
Dairy cows	400	93.2	60
Rustic heifers	600		} 58
Beef heifers	500		
Rustic cows	700	95.1	} 54
Beef cows	600	76.5	

The fertility expressed as calving rate at induced oestrus can be considered as good for bovine species. Progesterone plasma levels above 0.5 ng/ml ten days after implant removal is considered to indicate that females have a corpus luteum and have ovulated. The Table III shows also that rate of induction of ovulation is not the same according to the female group. In the rustic group, the percentage of females which ovulated after treatment is very close to 100p.cent, but around 25p.cent of the beef cows have no induction of ovulation.

Thus the response to hormonal treatments and management methods is not the same for all females in ovarian inactivity. Some have an ovulation induced, others have not. These results support the hypothesis that ovarian inactivity is not a steady state and one needs complementary definition of ovarian inactivity.

III.- DEFINITION OF OVARIAN INACTIVITY

In the ewe during post-partum and seasonal anoestrus, COGNIE et al. (1982), TERQUI and COGNIE (1984) described two main stages "deep" and "weak"inactivity and two transitory phases.

* "Deep" inactivity: low plasma FSH levels, the number of LH pulses between 2 to 4 in 12 hours, no oestradiol-17 (E_2) measurable in utero-ovarian vein and no E_2 response after a PMSG injection.

. Transitory phase: FSH is high, the number of LH pulses is similar, but E_2 is detected in utero-ovarian vein and increases after a PMSG injection.

* "Weak" inactivity: FSH levels are medium, the number of LH pulses is between 2 to 4, E_2 is secreted in large amounts and rises to very high levels in response to PMSG.

. Transitory phase: the number of LH pulses increases to 6 to 9, E_2 is very high and ovulation will occur.

In the cow during post-partum, weak ovarian inactivity and the last transitory phase have been described by various groups (TERQUI et al., 1982; SCHALLENBERGER et al., 1982; PETERS and LAMMING, 1984). Deep ovarian activity i.e. low FSH and low LH secretion is observed in underfed cows (GAUTHIER et al., 1983; GAUTHIER et al., 1984).

TABLE IV Endocrine definition of ovarian inactivity in ewes and cows

			LH	FSH
DEEP				
	FSH+ and LH-			
		ewe (1)	0.75 pulse/6h	2.6
		cow (2)	1.9 ng/ml	86
WEAK				
	FSH + and LH-			
		ewe (1)	0.75 pulse/h	3.7
		cow (2)	2.5 ng/ml	98

(1) From OUSSAID, 1983;   (2) From GAUTHIER et al., 1983.

In Table 4, FSH and LH secretion in weak and deep ovarian inactivity are presented for ewes and cows.

Furthermore, in the ewe OUSSAID (1983) has shown that in deep ovarian inactivity (low FSH and low LH), both FSH and LH have to be injected to induce ovulation but, during week inactivity when LH is only low, pulsatile injection of LH is able to induce ovulation. In the cow, such relationship between hormone plasma levels and hormones needed for induction of ovulation is not yet established.

CONCLUSION

Post-partum ovarian inactivity in cows could be divided into two main situations. Treatments like GnRH injections, temporary calf removal, introduction of the bull are efficient to induce ovulation in female in weak inactivity but not in females in deep inactivity when both FSH and LH are required. The variable percentage of females in deep ovarian inactivity between experiments presumably explains, for example, the variable results obtained after GnRH injection by different groups (ROCHE et al., 1982; PETERS and RILEY, 1982; ROCHE and IRELAND, 1984).

However, the calving rate and the percentage of induction of ovulation observed after treatment clearly indicate that the non induction of oestrus and ovulation is not the only problem of reproduction during post-partum but other aspects such as quality of females and male gametes have also to be taken into account.

REFERENCES

Cognié, Y., Gayerie, F., Poulin, N. and Saumande, J. 1982. Ovarian pituitary dialogue during the post-partum period in the ewe. Curr. Top. Vet. Med. Anim. Sci., 20, 305-313 (Ed. H. Karg and E. Schallenberger). (Martinus Nijhoff).
Gauthier, D., Terqui, M. and Mauléon, P. 1983. Influence of nutrition on pre-partum plasma levels of progesterone and total oestrogens and post-partum plasma levels of luteinizing hormones and follicles stimulating hormones in suckling cows. Anim. Prod., 37, 89-96.
Gauthier, D., Petit, M., Terqui, M. and Mauléon, P. 1984. Undernutrition and Fertility. "Les colloques de l'INRA", 27, 105-124. (Ed. R. Ortavant et H. Schindler). (INRA).
Monje, A., Alberio, R., Schiersmann, G., Chedrese, J. and Carou, N. 1983. Efecto de la prensencia del Macho Sobre la actividad sexual posparto de vacas de cria en dos niveles nutritionales. Abst. IX Reunion Latino Americana de Produccion Animal. Santiago Chile.
Oussaid, B. 1983. Etude de l'activité ovarienne et de la stimulation pendant l'anoestrus saisonnier chez la Brebis Ile-de-France. Thèse 3ème cycle. Univ. Paris VI, 39 p.
Pelot, J., de Fontaubert, Y., Chupin, D. and Terqui, M. 1984. Managment of reproduction in cattle: ovarian activity hormonal treatments and fertility. "Les colloques de l'INRA", 27, 55-70 (Ed. R. Ortavant et H. Schindler). (INRA).
Peters, A.R. and Riley, G.M. 1982. Pulsatile LH secretion and its induction in post-partum beef cows. Curr. Top. Vet. Med. Anim. Sci., 20, 225-228 (Ed. H. Kargh and E. Schallenberger). (Martinus Nijhoff).
Peters, A.R. and Lamming, G.E. 1984. Endocrine changes in the post-partum period. 10th Int. Congr. Anim. Reprod. Artif. Insem., IV, III 17-III 24.
Roche, J.F., Edwards, S. and Niswender, G.D. 1982. Induction of ovulation in post-partum beef cows. Curr. Top. Vet. Med. Anim. Sci., 20, 225-228 (Ed. H. Karg and E. Schallenberger). (Martinus Nijhoff).

Roche, J.F. and Ireland, J.J. 1984. Manipulation of ovulation in cattle. 10th. Int. Cong. Anim. Reprod. Artif. Insem., IV, IV 9-IV 17.

Schallenberger, E., Oerterer, U. and Hutterer, G. 1982. Neuroendocrine regulation of post-partum function. Curr. Top. Vet. Med. Anim. Sci., 20, 123-147. (Ed. H. Karg and E. Schallenberger). (Martinus Nijhoff).

Terqui, M., Chupin, D., Gauthier, D., Pelot, J. and Mauléon, P. 1982. Influence of managment and nutrition on post-partum endocrine function and ovarian activity in cows. Curr. Top. Med. Anim. Sci., 20, 384-408. (Ed. H. Karg and E. Schallenberger). (Martinus Nijhoff).

Terqui, M. and Cognié, Y. 1984. Definition of ovarian activity and restiration of pituitary and ovarian functions in ewes and cows. Les colloques de l'INRA, 27, 11-23 (Ed. R. Ortavant et H. Schindler). (INRA).

Terqui, M. and Legault, C. 1984. Reproductive potential in females. 10th Int. Cong. Anim. Reprod. Artif. Insem., IV, X 16-X 25.

Thatcher, W.W., Guilbault, L.A., Collier, R.J., Lewis S., Drost M., Knilkerbocker, J., Foster, D.B. and Wilcox, D.J. 1982. The impact of ante-partum Physiology on Postpartum performance in cows. Curr. Top. Vet. Med. Anim. Sci., 20, 1-25 (Ed. H. Karg, E. Schallenberger). (Martinus Nijhoff).

ENDOCRINE MECHANISMS CONTRIBUTING TO POSTPARTUM ANOESTRUS IN DAIRY AND BEEF CATTLE

E. Schallenberger and D.L. Walters*)

Institut für Physiologie, Südd. Versuchs- und Forschungsanstalt
für Milchwirtschaft, Technische Universität München,
8050 Freising-Weihenstephan
Federal Republic of Germany

*) M.R.C. Group in Reproductive Biology, University of Western Ontario,
University Hospital, London, Ontario N6A 5A5
Canada

ABSTRACT

The neutral stimuli evoked by sucking a beef cow are impinging upon hypothalamic function and prevent for a certain period the increase in frequency of discontinuous GnRH release necessary for unimpaired pituitary function. In addition, a concomitant hyperadrenal status is able to suppress the GnRH mediated LH release from the pituitary gland. Removal of the suckling stimulus early post partum or the "escape" from its inhibitory influence at varying times pp results in increased endogenous GnRH supply stimulating the accumulation of releaseable pools first of FSH and thereafter of LH. The same cascade of events occurs about 4-6 weeks earlier in milked dairy cows in which about "normal" pituitary gonadotrophin secretion can be achieved as early as Days 10-15 pp.

The gradual increase in pulsatile LH and FSH release pp induces an augmentation of LH and FSH receptors in large ovarian follicles. But a certain refractoriness to physiologic gonadotrophin stimulation lasts for about 14 days and oestradiol secretion remains still reduced thereafter. The pituitary gland is subject early pp to the negative feedback of ovarian oestradiol-17β. This prevents a too fast increase of frequency of LH but not FSH secretory bursts.

The ability of oestradiol-17β to exert the positive feedback on LH is recovering by 10 days pp in dairy as well as beef cows. Two types of preovulatory-like gonadotrophin releases can be induced after administration of oestradiol: high amplitude LH and concomitant FSH surges usually accompanied by visible signs of oestrus and low amplitude LH surges without any parallel FSH increase and often without symptoms of oestrus. Also, spontaneous preovulatory LH releases early pp are often of low magnitude and without symptoms of heat. The frequency of pulsatile LH and FSH secretion during this first periovulatory phase pp is lower than during regular oestrus.

Administration of GnRH either in a repeated discontinuous fashion or as constant infusion can stimulate as early as Day 5 pp sufficient pituitary function to allow oestradiol to evoke its positive feedback action. The resulting luteal progesterone secretion is impaired. LH but not FSH pulse frequency gets more reduced during the first than during the succeeding luteal phases. These diminished luteotrophic stimuli might be one factor amongst others such as high

uterine prostaglandin concentrations or hyperadrenal status to contribute to the so called "short cycle syndrome" evident post partum.
    The influence of the uterus of the dairy cow upon the resumption
of cyclicity is not understood. The frequent suckling induced oxytocin stimuli facilitating first uterine constriction and then involution are missing. Ovarian function does not resume until uterine
prostaglandin. F-2a concentrations have decreased below a certain
threshold limit what occurs again by about 2 weeks pp. Hysterectomy at
parturition results in an immediate cessation of PGF-2a production
and an increase of frequency and amplitude of pituitary LH but not FSH
secretion. The mechanisms of this interaction are yet unknown. Contrary to this selective pituitary response, the ovary is unable to
secrete earlier or more steroids in response to this greater gonadotrophic stimulation. Progesterone released by the persisting corpus
luteum after hysterectomy does not exert a negative feedback on LH
during an initial 10 days period.
    Ovariectomy pp results in an immediate increase of both LH and
FSH again demonstrating that the hypothalamo-pituitary axis of post-
partum dairy cows is under an inhibitory control which is depending
amongst other influences on an intact ovarian-uterine unit. It remains
to be proven if these latter mechanisms are also effective in the beef
cow in which suckling per se causes already a certain minimum period
of pp acyclia.

INTRODUCTION

    This contribution does not intend to repeat the detailed reviews
about postpartum reproductive performance of cattle (e.g. Wagner and
Oxenreider, 1971, Lamming, 1978, Carruthers et al., 1980, Karg,
1981, Lamming, Wathes and Peters, 1981, Schallenberger, Oerterer and
Hutterer, 1982, Haresign, Foxcroft and Lamming, 1983, Karg and
Schallenberger, 1983, Dziuk and Bellows, 1983, Peters and Lamming,
1984, Schallenberger et al., 1984), but rather tries to add some
additional aspects to the topic.

    In order to resume fertility post partum the following recovery
processes have to be accomplished in a sequential manner (Malven,
1984): (1) recovery from pregnancy state, (2) escape from
(suckling)-induced inhibition of gonadotrophins, (3) initiation of
luteal development with or without preceding ovulation, (4) occurrence
of oestrus with ovulation and (5) luteal life span sufficient to allow
maternal recognition of pregnancy. Various environmental, genetic and
management influences as well as diseases of the individual animal may
superimpose these processes and may prolong the interval until
recovery of cyclicity (for review see Terqui et al., 1982). Frequency
and intensity of suckling and minimal nutritional status exert thresh-

old effects (for review see Lamming, 1978; Short et al., 1972; Wiltbank and Cook, 1958). Ovulation can occur when animals are able to "escape" from the inhibitory influence of suckling (Hammons et al., 1973) or when they have passed a certain minimum acyclic period. The relative importance of one or a combination of various of these above mentioned factors has to be elucidated. The mechanisms, how these effects can impinge upon the endocrine ystem are yet unclear.

In order to gain some insight into the resumption of cyclicity pp the various levels of interactions have to be addressed in most standardized experimental conditions specifically examining the function of the hypothalamo-pituitary axis, the ovarian responsiveness and the contribution of the uterus. The suckled beef or the milked dairy cow have to be considered separately as the restoration of the hypothalamo-pituitary function leading to fertility pp seems to be postponed long enough in suckler cows to potentially override the problems caused by impaired ovarian responsiveness to gonadotrophins as well as a not yet fully involved uterus.

Pituitary gonadotrophin content and release:

Pituitary LH content increases from a nadir at parturition until Days 10-20 pp whereas pituitary FSH content is highest at parturition and decreases thereafter (Labhsetwar et al., 1964; Wagner, Saatman and Hansel, 1969). The milked dairy cow releases after single injections of high doses of GnRH about unchanged FSH throughout the postpartum period (Schallenberger, Schams and Zottmeier, 1978; Foster, Lamming and Peters, 1980) whereas LH release is considerably suppressed during the first week pp (Kesler et al., 1977; Fernandes et al., 1978; Schallenberger et al., 1978), contrary to the suckled beef cow in which pituitary LH release is weak for further 10-20 days (Webb et al., 1977). The longest recovery period is required after autumn calvings.

Removal of the suckling stimulus in beef cows at 3 weeks pp results in an immediate increase in frequency of spontaneous LH release, number of follicular LH receptors and follicular fluid prolactin content (Walters et al., 1982a). Pituitary LH and FSH content was not different between suckled and weaned beef cows 3 weeks pp (Walters et al., 1982b) but GnRH induced LH release in vitro was

greater in pituitary explants of weaned compared to previously
suckled cows. Obviously the pituitary gland of the latter animals had
experienced a certain lack of GnRH priming.

FSH secretion is not considered a factor limiting restoration of
cyclicity pp in milked as well as suckled cows. FSH plasma concentrations increase within the first few days pp (Schams et al., 1978;
Foster et al., 1980; Webb et al., 1980) revealing a negative
correlation with the length of the postpartum acyclic period in
suckled beef cows (Peters and Lamming, 1984).

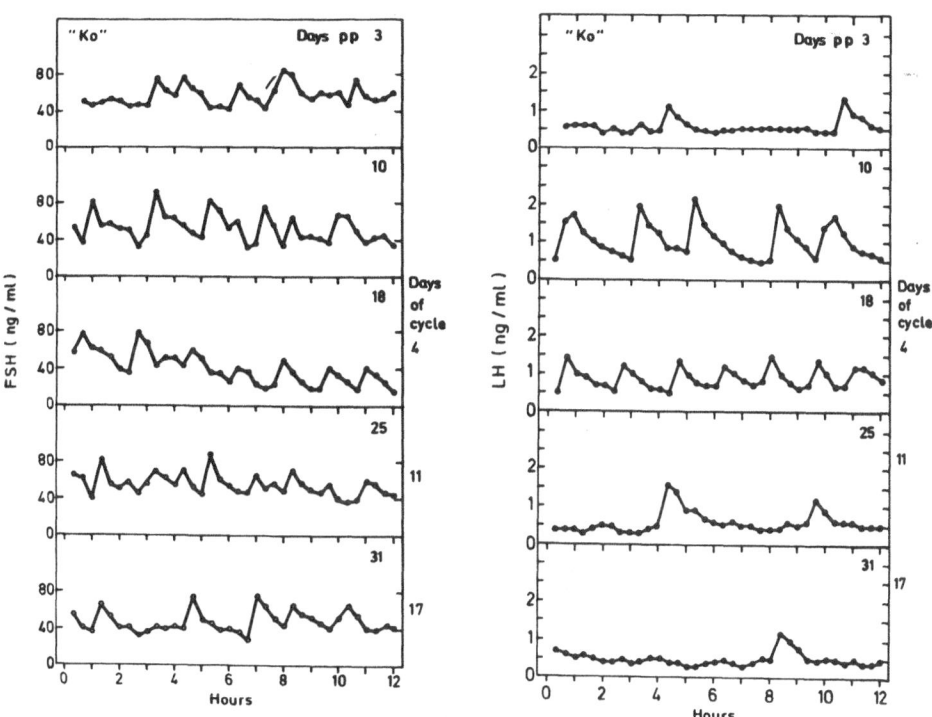

FIG. 1:   FIG. 2:
Pulsatile FSH and LH secretion characterized in about weekly intervals
pp. Blood was sampled every 20 min for 12 h. Oestrus was exhibited at
Day 35 pp; progesterone monitoring allowed to determine a 21 day cycle
with previous silent heat as indicated. (From Schallenberger, 1985).

There is always more frequent discontinuous FSH (FIG. 1) (average
frequency 1 pulse/2 h) than LH release (FIG. 2) present early pp.
Suckled beef cows have a postponed onset of short-term LH secretion

210

compared to suckled dairy or milked cows (Lamming et al., 1982). The frequencies of short-term secretion of LH increase in dairy cows during varying time periods (1.5 - 5 weeks pp) from about 1 pulse/4 h to 1 pulse/h prior to the reestablishment of luteal function, a period when frequencies of LH and FSH become identical (Schallenberger, 1985) (FIG. 6). Nevertheless interpulse intervals remain longer than during the periovulatory period of regular oestrous cycles (Walters and Schallenberger, 1984).

The magnitude of the first preovulatory LH surges pp is generally low (Schams et al., 1978; Lamming et al., 1981) and lacking a concomitant preovulatory FSH increase (FIG. 3). Visible symptoms of heat are often not detectable during these first periovulatory periods pp. If one agrees that FSH as well as LH secretion is triggered by discontinuous GnRH stimulation, one has to assume that the hypothalamus of the parturient dairy cow is continuing to supply the pituitary gland with GnRH stimuli. The divergency of both gonadotrophins might be explained by the fact that a releaseable pool of LH has to be built up first and that LH secretion might be subject in addition to a specific slowly vanishing inhibitory influence early pp. The GnRH mediated LH release might be in addition suppressed by high corticoid concentrations early pp (Li and Wagner, 1983b).

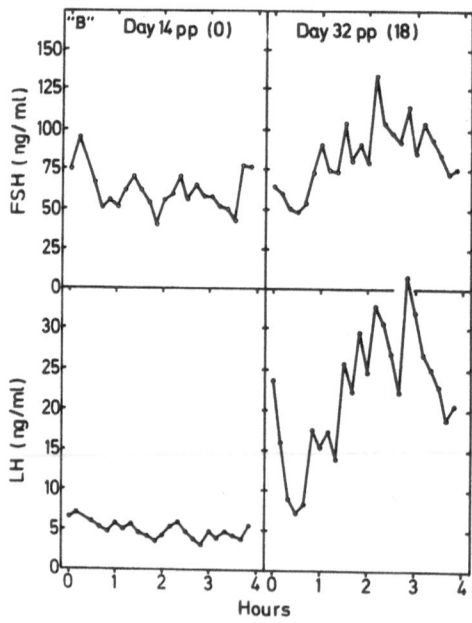

FIG. 3: Four h windows of periovulatory gonadotrophin surges of one cow at Day 14 (silent heat) and 32 (visible heat) pp. Note the low magnitude LH increase at Day 14 and the high magnitude LH/FSH amplitudes at Day 32 pp.

GnRH administration and pituitary function:

Substituting postpartum beef or dairy cows with repeated low dose discontinuous GnRH administration (Riley, Peters and Lamming, 1981; Walters et al., 1982c; Schallenberger et al., 1982; Edwards, Roche and Niswender, 1983) induces pulsatile LH and FSH releases, but there is variable success regarding ovulation. This might depend on the used frequency on GnRH injections (often 1 pulse/2 h) the used dose (often inducing higher LH magnitudes than seen in the postpartum cow) and the time and duration of the GnRH application. Also continuous GnRH infusion results in increased LH pp (Haresign, Foxcroft and Lamming, 1983).

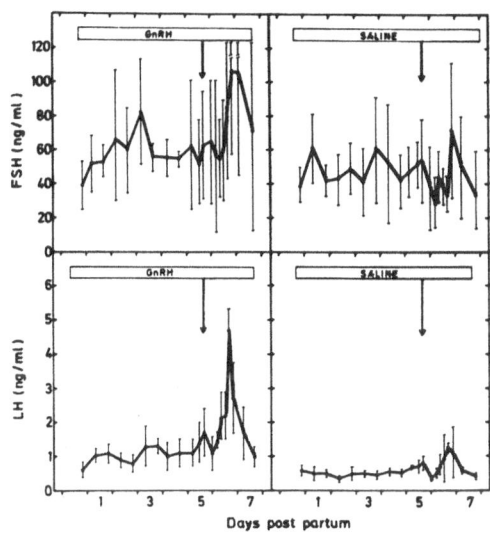

FIG. 4: Average LH and FSH concentrations ($\pm$ SD) of 4 GnRH treated (left panel; 1 pulse of 500 ng/h and 2 saline treated (right panel) heifers. ⬇ im injection of 1 mg oestradiol-17β-benzoate.
(From Schallenberger et al., 1984).

When giving GnRH with the approximate preovulatory frequency (1 pulse/h) from parturition throughout the first week pp we could induce in 4 out of 6 cows a gradual increase of mean plasma LH and FSH concentrations compared to controls (FIG. 4) as well as higher amplitudes and more frequent LH secretion by Day 5 compared to Day 1 pp (Schallenberger, Oerterer and Hutterer, 1982). Challenging these cows at Day 5 pp with oestradiol-17β benzoate did result in low magnitude preovulatory type LH and FSH surges (FIG. 4) and following luteal phase. The two animals with highest lactation failed to respond to the treatment (Karg and Schallenberger, 1983).

Resumption of the positive feedback:

The pituitary gland is not only depending on hypothalamic stimulation but is most suspectible to the feedback control of ovarian
steroids. The recovery of the stimulatory ability of oestradiol-17β
(positive feedback) is a prerequisite for the initiation of cyclicity
(Short et al., 1979; Stevenson, Spire and Britt, 1983; Schallenber -
ger, Oerterer and Hutterer, 1982). The delay in the restoration of the
LH release mechanism is supported by the suckling and milking stimuli
(Short et al., 1979) and is also depending on the presence of the
ovaries (Stevenson et al., 1983). The failure to exert positive feedback on both gonadotrophins (FIG. 5) and to induce heat during the
early postpartum period is not depending on the dose of administered
oestradiol-17β. The reported inhibition of FSH (Butler et al., 1983)
occurred from parturition throughout the entire pp period (FIG. 5).
Stimulation of LH and FSH was possible as early as Day 10 pp in dairy
cows (FIG. 5), allowing to classify 2 types of responses (Schallenberger and Prokopp, 1985). Parallel high magnitude LH / high magnitude
FSH surges accompanied most often by heat symptoms occurred when
already relatively frequent periovulatory-like LH and FSH release was
present prior to the challenge. Low magnitude LH surges without concomitant FSH releases were induced when less frequent LH and FSH release was present. Visible symptoms of heat were often missing at these
incidences. Therefore, the positive feedback mechanism is not re-
initiated spontaneously but needs a certain minimal pituitary
function.

Early pp low ovarian oestradiol releases are necessary to allow an
appropriate pituitary priming. This can be simulated by implanting
subcutaneously oestradiol-17β releasing capsules (Walters et al.,
1982c), which initially suppress to a certain extent pulsatile
pituitary LH release but facilitate higher magnitude positive feedback
and normal duration oestrous cycles lateron. The time required until
expression of the LH maximum (18-79 h) after inserting oestradiol implants is shorter the more the pp period has progressed (Peters and
Lamming, 1984).

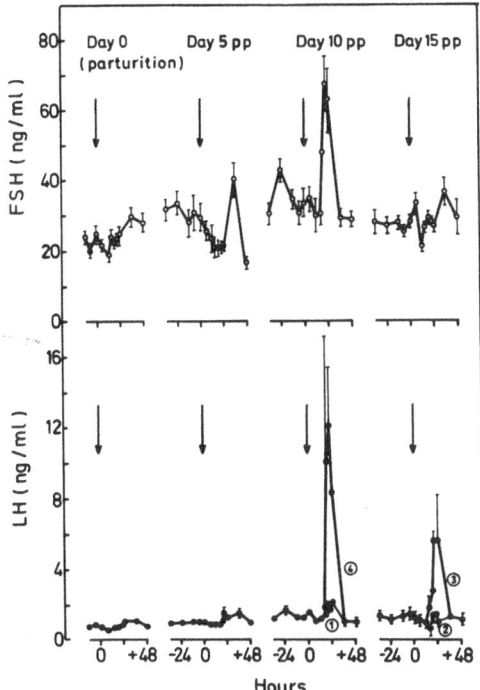

FIG. 5:
Reestablishment of the positive feedback action of oestradiol-17β-benzoate (↓ 1 mg im) given at Days 0, 5, 10 and 15 pp on LH and FSH (means ± SEM of 4 groups of 5 cows each). Note that 1 resp. 2 cows failed to exhibit LH surges at Days 10 and 15 pp. (Schallenberger and Prokopp, 1985).

Secretion of ovarian hormones pp:

Progesterone is the key hormone for the assessment of cyclicity pp and a wide range of information was obtained in the last decade by routine monitoring of progesterone profiles mainly derived from the analysis of milk samples (for reviews see Bulman and Lamming, 1978; Karg, 1981). Progesterone determination was since widely applied for detection of pathophysiological conditions and allowed a tremendous insight into the pattern of resumption of ovarian cyclicity proving that silent heat is most common pp.

Another key to resolve these problems is the measurement of ovarian oestradiol-17β. Only few relevant data are available. Concentrations in the peripheral plasma are low early pp. It is difficult to evaluate the minor differences to the luteal phase concentrations of regular cycles (Erb et al., 1971; Sasser et al., 1979). Pattern derived from milk samples (Pope, 1982) allowed the conclusion that there are irregular oestrogen increases already during the pp acyclic period representing spontaneous follicular growth with about normal oestrogen secretion. But there is an initial failure of positive feedback action of oestrogens at hypothalamus and pituitary gland even in the absence of high progesterone concentrations.

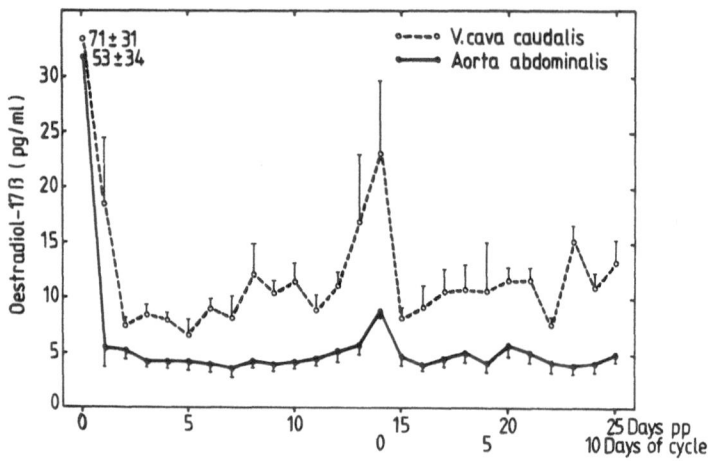

FIG. 6:
Average oestradiol-17β concentrations (± SEM) of 6 heifers throughout the first 25 Days pp; resumption of cyclicity was normalized to Day 14 pp. Samples were taken simultaneously from the vena cava caudalis and the aorta abdominalis.

Analysing oestradiol-17β obtained from samples collected from the aorta abdominalis and the vena cava caudalis cannulated according to the method of Walters, Schams and Schallenberger (1984) revealed slightly lower arterial oestradiol-17β concentration (FIG. 6) than present during regular cycles (for comparison see Walters and Schallenberger, 1984) The oestradiol levels in the vena cava caudalis representing the acute ovarian drainage indicate even better that the ovary is secreting impaired absolute amounts of oestrogens throughout the first weeks pp including the first periovulatory period. These concentrations are potentially not able to trigger the hypothalamo-pituitary system. This can be another reason for silent heat and impaired preovulatory-like gonadotrophin releases pp. Nevertheless, preovulatory oestradiol-17β is secreted in a pulsatile fashion revealing the same frequency as LH (FIG. 7). The typical oestradiol increase present during Days 3-5 of regular oestrous cycles (Walters, Schams and Schallenberger, 1984) is always missing during the first postpartum luteal phase (FIG. 6). Also, the early preovulatory follicle pp secretes already low amounts of oxytocin (FIG. 7) contrary to the missing oxytocin secretion of the periovulatory follicle during regular cycles (Walters and Schallenberger, 1984). Progesterone and oxytocin releases from the corpus luteum into the vena cava occur in a pulsatile fashion having the same frequency as FSH (FIG. 7). Average peripheral progesterone concentrations remain often lower early pp than during regular cycles. This could be also influenced by enhanced adrenal corticosteroid concentrations early pp as Li and Wagner (1983a) demonstrated a suppression of luteal progesterone synthesis after induced hyperadrenalism.

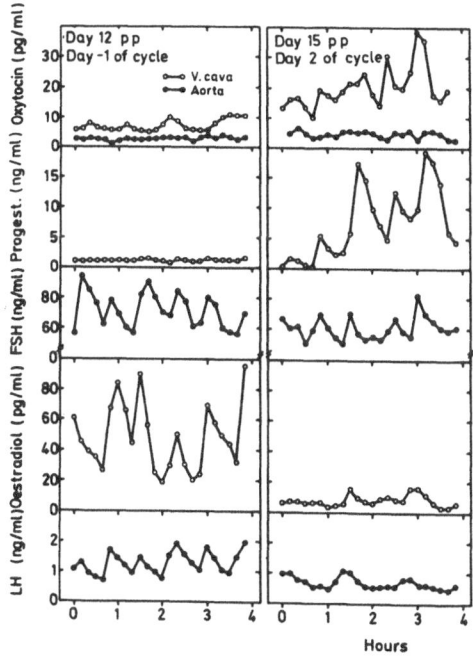

FIG. 7:
Depiction of 4 h windows taken at Days 12 and 15 pp representing cycle Days -1 and +2. Progesterone, oestradiol-17β and oxytocin were determined from blood samples collected every 10 min from the vena cava caudalis and LH and FSH from simultaneous samples taken from the aorta abdominalis.

Even when the ovary is secreting only basal oestradiol during the first days pp it seems to exert or transmit an inhibitory influence upon the pituitary gland. Ovariectomy performed at Day 4 pp results in an immediate increase in frequency as well as amplitude of FSH identical to the increase resulting after ovariectomy performed during

216

the non-luteal phase of the cycle (Schallenberger and Peterson, 1982). Frequency of LH release increases similarly but amplitude pp is some- what reduced for an about 2 week's duration (FIG. 8).

The role of the uterus:

The influence of the uterus on the resumption of cyclicity pp has received only few attention. It is known that the high concentrations of the major prostaglandin F-2α-metabolite (13,14-dihydro-15-keto-PGF-2α) secreted from the uterus early pp have to decrease below a certain lower threshold prior to initiation of cyclicity pp (for review see Kindahl et al., 1982; Thatcher et al., 1982).

Performing hysterectomy within 5 h after parturition (Schallen- berger et al., 1984) results within 30 minutes in a decrease of PGF-2α and PGFM (Meyer et al., 1984) to about a 1/10th of the control con- centrations and to about 1/40th within another 7 hours. These low con- centrations are not reached until about 14 days later in the control animals. This approach uncovered an inhibitory influence of the uterus selectively acting on the LH but not FSH secretion. Pituitary LH release increased immediately after surgery (FIG. 8) in an almost similar fashion as if the animals would have been ovariectomized (FIG. 8) with the exception that FSH release remined unaltered.

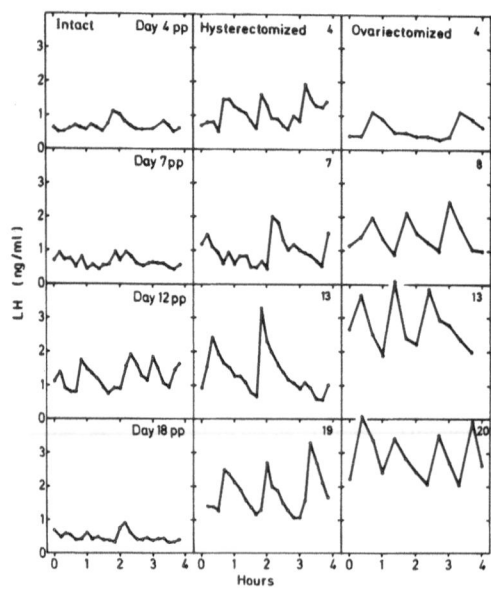

FIG. 8:
Pulsatile LH secretion post partum.
Left panel: intact animal with resumption of luteal function at Day 14 pp.
Middle panel: animal hyster- ectomized within 5 h pp.
Right panel: animal ovariectomized at Day 4 pp. Samples were collected every 10-20 mins for 4 h.

Even in the presence of these enhanced LH concentrations the ovary did not resume activity earlier nor did it produce higher oestradiol and progesterone concentrations (Schallenberger et al., 1984) again pointing towards a certain ovarian refractoriness to gonadotrophic stimulation for a minimum period pp. Progesterone secretion from the persisting corpus luteum seemed to increase after an initial 2 week's period of lower concentrations. Also, despite progesterone secretion sufficient to induce an immediate negative feedback on LH in intact animals (FIG. 8), the enhanced LH concentrations in the hysterectomized cows were not subject to the negative feedback action until after the initial 10 days of luteal phase.

OUTLOOK

These finding support the idea that at least the dairy cow which is neither suffering from a strong nor an extended hypothalamo-pituitary quiescence has developed alternative mechanisms involving ovary and uterus to prevent the animal from too early rebreeding. The beef cow has sufficient time to recover from pregnancy.

Early resumption of cyclicity pp was a goal of cattle breeding and management missing sometimes the fact that there was a tremendous gap between the time of initiation of cyclicity and fertility. Studies of the basic endocrine mechanisms might contribute to a better understanding of these problems. Various approaches (biotechnical measures, breeding, husbandry) might help to achieve more synchrony of the reinitiation of cyclicity and fertility and thus might improve reproductive performance of cattle.

REFERENCES

Bulman, D.C. and Lamming, G.E. 1978. Milk progesterone levels in relation to conception, repeat breeding and factors influencing acyclicity in dairy cows. J. Reprod. Fertil. 54, 447-458.

Butler, W.R., Katz, L.S., Arriola, J., Milvae, R.A. and Foote, R.H. 1983. On the negative feedback regulation of gonadotropins in castrate and intact cattle with comparison of two FSH radioimmunoassays. J. Anim. Sci. 56, 919-929.

Carruthers, T.D., Convey, E.M., Kesner, J.S., Hafs, H.D. and Cheng, K.W. 1980. The hypothalamo-pituitary gonadotrophic axis of suckled and nonsuckled dairy cows postpartum. J.Anim.Sci. 51, 949-957.

Dziuk, P.J. and Bellows, R.A. 1983. Management of reproduction of beef cattle, sheep and pigs. 1983. J. Anim. Sci. 57, Suppl. 2, 355-379.

218

Edwards, S., Roche, J.F. and Niswender, G.D. 1983. Response of suckling beef cows to multiple, low-dose injections of Gn-RH with or without progesterone pretreatment. J. Reprod. Fertil. 69, 65-72.

Erb, R.E., Surve, A.H., Callahan, C.J., Randel, R.D. and Garverick, H.A. 1971. Reproductive steroids in the bovine: changes post-partum. J. Anim. Sci. 33, 1060-1071.

Fernandes, L.C., Thatcher, W.W., Wilcox, C.J. and Call, E.P. 1978. LH release in response to GnRH during the postpartum period of dairy cows. J. Anim. Sci. 46, 443-448.

Foster, J.P., Lamming, G.E. and Peters, A.R. 1980. Short-term re-lationships between plasma LH, FSH and progesterone concen-trations in post-partum dairy cows and the effect of Gn-RH in-jection. 1980. J. Reprod. Fertil. 59, 321-327.

Hammons, J-A., Velasco, M. and Rothchild, I. 1973. Effect of the sudden withdrawl or increase of suckling on serum LH levels in ovariectomized postparturient rats. Endocrinology 92, 206-211.

Haresign, W., Foxcroft, G.R. and Lamming, G.E. 1983. Control of ovu-lation in farm animals. J. Reprod. Fertil. 69, 383-395.

Karg, H. 1981. Physiological impact on fertility in cattle with special emphasis on assessment of the reproductive function by progesterone assay. Livestock Prod. Sci. 8, 233-246.

Karg, H. and Schallenberger E. 1983. Regulation der ovariellen Steroid-hormonkretion post partum. Wien. tierärztl. Mschr. 70, 238-243.

Kesler, D.J., Garverick, H.A. Youngquist, R.S., Elmore, R.G. and Bier-schwal, C.J. 1977. Effect of days postpartum and endogenous re-productive hormones on GnRH-induced LH release in dairy cows. J. Anim. Sci. 46, 797-803.

Kindahl, H., Edqvist, L.-E., Larsson, K. and Malmqvist, A. 1982. In-fluence of prostaglandins on ovarian function post partum. In: Factors influencing fertility in the postpartum cow (Current Topics in Veterinary Medicine and Animal Science) Vol. 20, Eds. H.Karg and E. Schallenberger. Martinus Nijhoff, The Hague; pp 173-196.

Labhsetwar, A.P., Collins, W.E., Tyler, W.J. and Casida, L.E. 1964. Some pituitary-ovarian relationships in the periparturient cow. J. Reprod. Fertil. 8, 85-90.

Lamming, G.E. 1978. Reproduction during lactation. In: Control of ovu-lation. Eds. Crighton, D.B., Foxcroft, G.R., Haynes, N.B. and Lamming, G.E. Butterworths, London; pp 335-353.

Lamming, G.E., Wathes, D.C. and Peters, A.R. 1981. Endocrine patterns of the post partum cow. J. Reprod. Fertil. 30, 155-170.

Lamming, G.E., Peters, A.R., Riley, G.M. and Fischer, M.W. 1982. Endo-crine regulation of post-partum function. In: Factors influencing fertility in the postpartum cow (Current Topics in Veterinary Medicine and Animal Science) Vol. 20, Eds. H. Karg and E. Schallenberger. Martinus Nijhoff, The Hague; pp 148-172.

Li, P.S. and Wagner, W.C. 1983a. Effects of hyperadrenal states on lu-teinizing hormone in cattle. Biol. Reprod. 29, 11-24.

Li, P.S. and Wagner, W.C. 1983b. In vivo and in vitro studies on the effect of adrenocorticotropic hormone or cortisol on the pituitary response to gonadotropin releasing hormone. Biol. Reprod. 29, 25-37.

Malven, P.V. 1984. Pathophysiology of the puerperium: definition of the problem. Proc. 10th Int. Congr. Animal Reprod. & A.I., Urbana-Champaign (USA), Vol. IV, III.1-8.

Meyer, H.H.D., Schallenberger, E., Enzenhöfer, G. 1984. 13,14-dihydro-15-keto-prostaglandin $F_{2\alpha}$ determination in plasma of postpartum cows to monitor prostaglandin $F_{2\alpha}$ secretion. Acta Endocr. (Copenh.) Suppl. 264, 72-73.

Peters, A.R. and Lamming, G.E. 1984. Endocrine changes in the postpartum period. Proc. 10th Intern. Congr. Animal Reprod. & A.I., Urbana-Champaign (USA), Vol. IV, III.17-24.

Pope, G.S. 1982. Oestrogens and progesterone in plasma and milk of post-partum dairy cattle. In: Factors influencing fertility in the postpartum cow (Current Topics in Veterinary Medicine and Animal Science) Vol. 20, Eds. H. Karg and E. Schallenberger. Martinus Nijhoff, The Hague; pp 248-276.

Riley, G.M., Peters, A.R. and Lamming, G.E. 1981. Induction of pulsatile LH release, FSH release and ovulation in post-partum acyclic beef cows by repeated small doses of Gn-RH. J. Reprod. Fertil. 63, 559-565.

Sasser, R.G., Falk, D.E. and Ross, R.H. 1979. Estrogen in plasma of parturient paretic and normal cows. J. Dairy Sci. 62, 551-556.

Schallenberger, E., Schams, D. and Zottmeier, K. 1978. Response of lutropin (LH) and follitropin (FSH) to the administration of gonadoliberin (GnRH) in pregnant and post-partum cattle including experiments with prolactin suppression. Theriogenology 10, 35-53.

Schallenberger, E., Peterson, A.J. 1982. Effect of ovariectomy on tonic gonadotrophin secretion in cyclic and post-partum dairy cows. J. Reprod. Fertil. 64, 47-52.

Schallenberger, E., Oerterer, U. and Hutterer, G. 1982. Neuroendocrine regulation of postpartum function. In: Factors influencing fertility in the postpartum cow (Current Topics in Veterinary Medicine and Animal Science) Vol. 20, Eds. H. Karg and E. Schallenberger. Martinus Nijhoff, The Hague; pp 123-147.

Schallenberger, E., Walters, D.L., Oschmann, S.J. and Meyer, H.H.D. 1984. Endocrine changes during the early postpartum period in dairy cattle. Proc. 10th Intern. Congr. Animal Reprod. & A.I., Urbana-Champaign (USA), Vol. IV, III.9-16.

Schallenberger, E. 1985. Gonadotrophins and ovarian steroids in cattle. III. Pulsatile changes of gonadotrophin concentrations in the jugular vein post partum. Acta Endocrinol. (Copenh.) submitted for publication.

Schallenberger, E. and Prokopp, S. 1985. Gonadotrophins and ovarian steroids in cattle. IV. Reestablishment of the stimulatory feedback action of oestradiol-17β on LH and FSH. Acta Endocrinol. (Copenh.) submitted for publication.

Schams, D., Schallenberger, E., Menzer, Ch., Stangl, J., Zottmeier, K., Hoffmann, B. and Karg, H. 1978. Profiles of LH, FSH and progesterone in post-partum dairy cows and their relationship to the commencement of cyclic functions. Theriogenology 10, 453-468.

Short, R.E., Bellows, R.A., Moody, E.L. and Howland, B.E. 1972. Effects of suckling and mastectomy on bovine postpartum reproduction. J. Anim. Sci. 34, 70-74.

Short, R.E., Randel, R.D., Staigmiller, R.B. and Bellows, R.A. 1979. Factors affecting estrogen-induced LH release in the cow. Biol. Reprod. 21, 683-689.

Stevenson, J.S., Spire, M.F. and Britt, J.H. 1983. Influence of the ovary on estradiol-induced luteinizing hormone release in postpartum milked and suckled Holstein cows. J. Anim. Sci. 57, 692-698.

Terqui, M., Chupin, D., Gauthier, D., Perez, N., Pelot, J. and Mauléon, P. 1982. Influence of management and nutrition on postpartum endocrine function and ovarian activity in cows. In: Factors influencing fertility in the postpartum cow (Current Topics in Veterinary Medicine and Animal Science) Vol. 20, Eds. H. Karg and E. Schallenberger. Martinus Nijhoff, The Hague; pp 384-408.

Thatcher, W.W., Guilbault, L.A., Collier, R.J., Lewis, G.S., Drost, M., Knickerbocker, J., Foster, D.B. and Wilcox, C.J. 1982. The impact of ante-partum physiology on postpartum performance in cows. In: Factors influencing fertility in the postpartum cow (Current Topics in Veterinary Medicine and Animal Science) Vol. 20, Eds. H. Karg and E. Schallenberger. Martinus Nijhoff, The Hague; pp 3-25.

Wagner, W.C., Saatman, R. and Hansel, W. 1969. Reproductive physiology of the post-partum cow. II. Pituitary, adrenal and thyroid function. J. Reprod. Fertil. 18, 501-508.

Wagner, W.C. and Oxenreider, S.L. 1971. Endocrine physiology following parturition. J. Anim. Sci. 32, Suppl. I, 1-16.

Walters, D.L., Kaltenbach, C.C., Dunn, T.G. and Short, R.E. 1982a. Pituitary and ovarian function in postpartum beef cows. I. Effect of suckling on serum and follicular fluid hormones and follicular gonadotropin receptors. Biol. Reprod. 26, 640-646.

Walters, D.L., Short, R.E., Convey, E.M., Staigmiller, R.B., Dunn, T.G. and Kaltenbach, C.C. 1982b. Pituitary and ovarian function in postpartum beef cows. II. Endocrine changes prior to ovulation in suckled and nonsuckled postpartum cows compared to cycling cows. Biol. Reprod. 26, 647-654.

Walters, D.L., Short, R.E., Convey, E.M., Staigmiller, R.B., Dunn, T.G. and Kaltenbach, C.C. 1982c. Pituitary and ovarian function in postpartum beef cows. III. Induction of estrus, ovulation and luteal function with intermittent small-dose injections of GnRH. Biol. Reprod. 26, 655-662.

Walters, D.L. and Schallenberger, E. 1984. Pulsatile secretion of gonadotrophins, ovarian steroids and ovarian oxytocin during the periovulatory phase of the oestrous cycle in the cow. J. Reprod. Fertil. 71, 503-512.

Walters, D.L., Schams, D. and Schallenberger, E. 1984. Pulsatile secretion of gonadotrophins, ovarian steroids and ovarian oxytocin during the luteal phase of the oestrous cycle in the cow. J. Reprod. Fertil. 71, 479-491.

Webb, R., Lamming, G.E., Haynes, N.B., Hafs, H.D. and Manns, J.G. 1977. Response of cyclic and post-partum suckled cows to injections of synthetic LH-RH. J. Reprod. Fertil. 50, 203-210.

Webb, R., Lamming, G.E., Haynes, N.B. and Foxcroft, G.R. 1980. Plasma progestereone and gonadotrophin concentrations and ovarian activity in post-partum dairy cows. J. Reprod. Fertil. 59, 133-143.

Wiltbank, J.N. and Cook, A.C. 1958. The comparative reproductive performance of nursed and milked cows. J. Anim. Sci. 17, 640-648.

STUDIES ON THE ENDOCRINE AND ENVIRONMENTAL FACTORS INFLUENCING POST PARTUM ACYCLICITY IN DAIRY COWS BASED ON PROGESTERONE PROFILES (PRELIMINARY RESULTS)

J. Henriksen & A. Mikél Jensen*

Institute of Animal Science,
Royal Veterinary- and Agricultural University,
Rolighedsvej 23,
DK-1958 Copenhagen V, Denmark

*Department of Clinical Chemistry,
State Veterinary Serumlaboratory,
Bülowsvej 27,
DK-1870 Copenhagen V, Denmark.

ABSTRACT

The post partum period of 54 dairy cows was screened using progesterone determinations in fat free milk samples obtained every 2-3 days. Profiles were drawn of progesterone, milk yield and body weight change. On an average ($\pm$ s.d.) the cows had regained their early weight loss at 120 days post partum, the days open were 86 ± 27, and the period of acyclicity post partum was 33 ± 29 days. Cyclicity was resumed in 63% of the cows within 30 days and in 90% within 50 days post partum. First ovulation (after the first luteal phase) took place within 50 days post partum in 53% of the cows and within 70 days in 82%. A prolongation of the acyclic period was recorded together with a weight loss during the first 120 days post partum for some cows. Milk yield had no influence on the reproductive performance. The length of the post partum period of acyclicity was of importance for the onset of insemination, but had no decisive influence on the reproductive performance of the cows. Cows with an acyclic period of 30-60 days post partum could be inseminated and get pregnant within 90 days post partum.

222

INTRODUCTION

In Denmark, reproduction in dairy herds is often unsatisfactory, with too many days open, too many inseminations per cow, and many cows having to be culled because of fertility problems. An investigation carried out in 50 dairy herds with 3000 cows through the years 1979-1982 showed that management was the most important factor influencing reproduction. At the same time a great individual variability in reproductive performance was recorded (Henriksen et al., 1984). There was little evidence that this variability could be due to nutrition or milk yield, or to environmental factors, such as housing system. The period from calving to first insemination seemed to be of great importance on the number of the resulting days open. This means that the period between calving and resumption of the oestrous cycle could be a factor of decisive importance to the reproductive efficiency of the cow.

The objective of the present study was to investigate the post partum period in dairy cows and to see how this period influenced the reproductive performance. Of special interest was the acyclic period from calving to the first luteal phase. Variation in the length of the acyclic period between cows as influenced by body weight change, nutrition, and milk yield was elucidated.

The preliminary results of the investigation, primarily concerning the post partum period of acyclicity, are presented in this paper.

MATERIAL AND METHOD

The investigation was performed in a dairy herd of 54 Danish Holstein-Friesians, kept in a stanchion barn belonging to Silstrup Research Station. The cows were in a feeding trial in which they were offered the same amount of energy, protein, etc. The only difference between rations was in the composition of the carbohydrate fraction (digestible fiber, sugar, and starch). The feed uptake was recorded daily and the cows were weighed on day 4 after calving and once every fortnight thereafter. The milk yield was recorded twice weekly on two consecutive days and the milk was analysed for content of fat and protein.

Milk samples (10 ml) were collected twice weekly and used for progesterone analysis. The first sample was taken approximately 7 days after calving, and the last when new pregnancy was diagnosed or the cow culled. The milk samples were centrifuged and the progesterone content determined on fat-free milk by a direct radioimmunoassay technique (Oltner and Edqvist, 1980) as described by Ball and Pope (1976). The descriminatory level between the follicular phase and the luteal phase was 1.5 nmol/l; this implies that the progesterone level was below this value during the acyclic period post partum. Ovarian cyclicity was resumed when the level rose above 1.5 nmol/l, i.e. at the start of the first luteal phase. The ovulation taking place after the first luteal phase was considered to be the first ovulation, initiating the first normal oestrous cycle.

On account of the feeding trial, it was endeavoured that calvings should take place in September-November. Therefore inseminations did not start till late in November. With this limitation, the cows were inseminated in observed heat from day 35 post partum. The herdsman/technician was responsible for heat detection and insemination. The results of the progesterone analyses were available to the herdsman/technician about 2 weeks after collection of the milk samples.

About 30 days after calving the cows were examined by rectal palpation (for judgement of the state of the ovaries and of uterine involution) by an experienced veterinarian who also examined for pregnancy from day 40 after the last insemination.

The milk concentrations of progesterone allowed monitoring of the ovarian activity (Karg, 1981) without interfering with the cow at all, and without regard to management.

The data were analysed by a linear regression analysis.

RESULTS

Reproductive performance

Table 1 shows the means and variations of all parameters measured. The body weight change recorded at, respectively, 30, 60 and 120 days after calving was used as an expression of the energy-balance. On an average, the cows had compensated for earlier weight loss by 120 days post partum. Milk yield was measured as kg 4% fat corrected milk (FCM). As shown in Table 1, the cows were all relatively high yielders. The average reproduction result showed that the herd had a good reproductive performance. Days from first insemination to conception was calculated as zero for those cows which conceived at the first insemination.

TABLE 1. Means and variation (standard deviation = s.d.) of the parameters measured on the 54 cows.

Parameter	Mean	s.d.
Weight at calving	552 kg	74
Body weight change		
30 days post partum	− 18 kg	17
60 " " "	− 12 kg	20
120 " " "	+ 3 kg	32
Milk yield, kg FCM		
30 days post partum	884 kg	176
60 " " "	1750 kg	312
120 " " "	3352 kg	541
Days open (calving-conception)	86 days	27
First insemination - conception	18 days	21
Calving - first insemination	70 days	18
Acyclic period post partum	33 days	29

Table 2 shows that ovarian cyclic activity began before day 10 post partum in a few cows and between day 10 and day 19 post partum in one third of the cows. Cyclicity was resumed in 92% of the cows within 70 days post partum.

TABLE 2. The post partum acyclic period, defined as the period from calving to the first milk sample with a progesterone concentration above 1.5 nmol/l

	Days of acyclicity post partum							
	> 10	10-19	20-29	30-39	40-49	50-59	60-69	< 70	Total
Number of cows	3	17	14	10	3	1	1	5	54
%	6	31	26	19	6	2	2	8	100
Accumulated %	6	37	63	82	88	90	92	100	

Table 3 shows the mean number of days from calving to first ovulation. As shown in Table 4, ovulation occurred within an average of 15-25 days after the cycle had started. Table 5 shows the reproductive performance of the pregnant cows depending upon the length of the acyclic period post partum. The relatively long period from calving to first insemination was caused by the strict policy concerning calving season as mentioned earlier. The above mentioned parameters were responsible for only 18% of the variation in the length of the acyclic period post partum (Table 6).

TABLE 3. Interval from calving to first ovulation, defined as days from calving to the end of the first luteal-phase post partum.

		Days from calving to first ovulation								
	< 10	10-29	20-29	30-39	40-49	50-59	60-69	>70	Total	Mean
Number of cows	-	1	6	13	8	11	4	8	51	50
%	-	2	11	25	15	21	8	15	97	-
Accumulated %	0	2	13	38	53	74	82	97	-	-

TABLE 4. Interval from calving to first ovulation, as influenced by the length of the acyclic period post partum.

		Days of acyclicity post partum						
	<10	10-19	20-29	30-39	40-49	50-59	60-69	>70
Number of cows	3	17	14	10	3	1	1	3
Number of days:								
Calving to first ovulation	24	36	49	57	71	77	80	112
s.d.	6.8	9.5	10.6	11.5	5.3	-	-	12.1

Accumulated milk yield at 30, 60 and 120 days seemed to be of no significance in relation to post partum acyclicity (Table 6). Month of calving and body weight change did have some influence, especially since the accumulated weight loss at 120 days post partum seemed to be correlated with the length of the acyclic period. A further regression analysis showed that accumulated weight loss at 120 days post partum was very closely correlated with the accumulated milk yield at 30 days post partum.

The body weight change (weight loss) was correlated with both the period until first insemination and the days open; this was true especially of the weight loss at day 30 post partum, which delayed the onset of insemination.

Because of the insemination policy (see introduction) the month of calving had a very strong influence on the period from calving-to-first insemination and on days open. Table 6 also shows that a short period of acyclicity shortened the period to the first insemination, but had no significant effect either on the days from first insemination to conception or on the days open.

The other parameters examined seemed to have no effect. The multiple correlation coefficient (R^2) in Table 6 shows that only a smaller part of the variation could be attributed to these parameters.

TABLE 5. Reproductive performance of the pregnant cows* as influenced by the length of the acyclic period post partum (mean±standard error).

| Days of acyclicity post partum | Number of cows | Days open | Reproductive performance | |
			Days from calving to first insemination	Days from first insemination to conception
below 20	20	83±5.9	65±3.5	17±5.4
20–29	12	89±7.5	68±3.5	16±6.0
30–49	10	75±8.9	68±5.3	11±4.7
50–69	2	86	65	21

*two cows with embryonic loss were excluded

TABLE 6. Level of significance for the parameters for
reproduction performance in a linear regression
model

Parameter	Days of acyclicity post partum	Days from calving to first insemination	Days from first insemination to conception	Days open
Month of calving	0.05	0.01	N.S.	0.01
Parity	N.S.	N.S.	N.S.	N.S.
* Body weight change				
30 days post partum	0.07	0.05	N.S.	0.07
60 " " "	0.09	N.S.	N.S.	0.07
120 " " "	0.01	0.13	N.S.	0.15
Milk yield, kg CM				
30 days post partum	N.S.	N.S.	N.S.	N.S.
60 " " "	N.S.	N.S.	N.S.	N.S.
120 " " "	N.S.	N.S.	N.S.	N.S.
Days of acyclicity post partum	-	0.07	N.S.	N.S.
R^2	0.18	0.51	0.06	0.39

*) only one of the following parameters was taken into the model at a
time

N.S. = insignificant (P>0.20)

R^2 = the multiple correlation coefficient expresses the part of the
variation which is explained by the model

Progesterone profiles

The progesterone profiles of the individual cows demonstrated that 5 cows were acyclic for more than 70 days post
partum; 3 of these cows resumed ovarian cyclicity and were
inseminated, but did not conceive.

Prolonged luteal function with high progesterone level
for 25-46 days was seen in 7 cows. Three of the cows had
uterine infections. All the cows returned to normal by 80
days post partum, and pregnancy was established in 6 of
them.

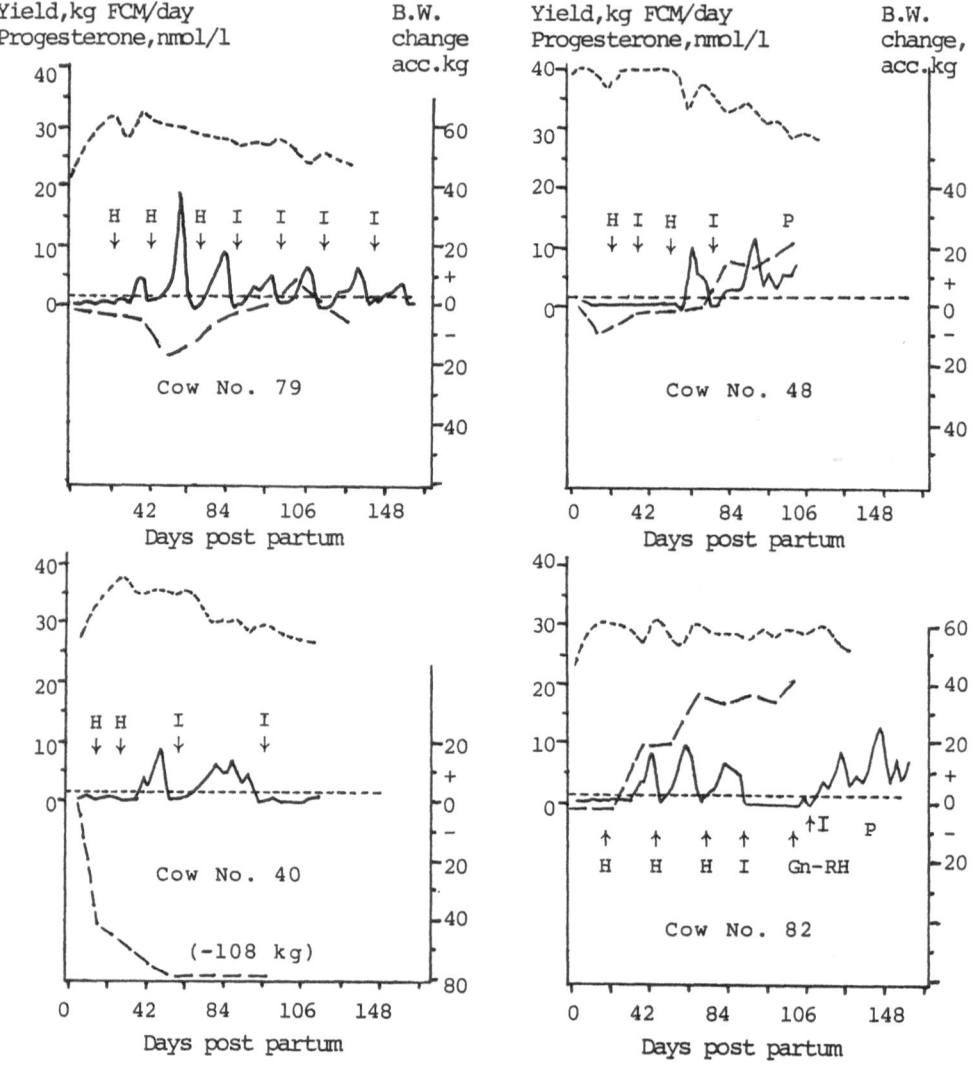

Figure 1. Profiles of milk yield (kg FCM per day), progesterone
nmol/l) and body weight change (accumulated kg) in
relation to days post partum
- - - milk yield, kg FCM per day
——— progesterone, nmol/l defatted milk
— — accumulated body weight change, kg
- - - -(horizontal line) 1.5 nmol/l, indicates descriminatory level between luteal and follicular phase
H = detected heat
I = insemination
Gn-RH = treatment with gonadotrophin releasing hormone
P = pregnant

Overall there was good correspondance between observed heat and progesterone levels below 1.5 nmol/l, and the number of inseminations per pregnancy was 1.9. Cystic ovaries were recorded in two cows, delayed ovulation in 3 cows and early embryonic death in 2 cows.

Figure 1 demonstrates progesterone profiles, milk yield, and body weight changes in 4 different cows. Cow No. 79 was observed in heat 21 days after calving, i.e., during the acyclic period, ovarian cyclicity was resumed 33 days post partum and the first luteal phase was short (about 7 days). Heat was thereafter observed at regular 3-week intervals, and insemination was started at the fourth heat. The first insemination was mis-timed, being carried out during prooestrus. After 3 further inseminations that did not result in pregnancy the cow was culled.

In cow No. 48 the acyclic period was 63 days. During this period signs of heat were observed 3 times, and insemination was performed at the second observed heat. The first luteal phase was short (10 days). Insemination at the following heat resulted in pregnancy. In cow No. 40 signs of heat were detected about 3 weeks after calving, but ovarian cyclicity was not resumed until 39 days post partum. The first luteal phase lasted for 14 days and the following insemination resulted in pregnancy, as confirmed by the high progesterone level. A rapid fall of that level 50 days later indicated embryo loss. The cow was re-inseminated but did not continue to cycle, and remained acyclic until culled. In cow No. 82, heat was observed 24 days post partum, the first luteal phase started at day 35 and lasted for 11 days. Signs of heat were recorded at regular 3 week intervals. Insemination was performed at the third heat, whereafter the cow became cystic and remained so for a long period. When oestrus was observed ten days after treatment with a gonadatrophin-releasing hormone, pregnancy was established following insemination.

DISCUSSION

Onset of cyclicity

The percentage of cows starting their oestrous cycle within 29 days post partum was 63% (Table 2), which is in good accordance to other findings for cows in a stanchion barn (Claus et al. 1982). Oestrus, with a possibility of successful insemination, occurred within 70 days post partum in 82% of the cows (Table 3). This is also in good agreement with the findings by Claus et al. (1982). It thus seems possible that insemination could be started early after calving, depending on oestrous behavior and efficiency of heat detection.

A short-term elevation in progesterone preceding the onset of normal ovarian and oestrous cycles was seen. Webb et al. (1980) discussed whether this first luteal phase was a result of ovulation or whether luteinized follicles were the source of the progesterone increase. In the present study oestrous behavior was seen during the acyclic period in about 35% of the cows, indicating that ovulation might occur in that period. However, the first ovulation was not reckoned to occur till after the first luteal phase.

Reproductive performance

The effect of negative energy balance on the reproductive performance was reviewed by Sejrsen and Neimann-Sørensen (1982). In agreement with their findings a prolongation of the acyclic period post partum was recorded in the present study when the total weight loss, especially at 120 days post partum, was great. The accumulated weight loss until day 120 post partum could be regarded as a measure of an energy deficit in the first part of lactation. There was a tendency for the length of the period from calving to first insemination, as well as for the number of days open, to increase with increasing weight loss (Table 6). On the other hand, unlike many other researchers (see Schmidt and Farreis, 1982), we found no influence of milk yield on the reproductive performance, although there was a strong correlation between the accumulated weight loss at 120 days post partum and the milk yield until 30 days. This observation indirectly supports the generel hypothesis that an early yield peak is unfavorable for fertility.

The influence of the length of the acyclic period on the reproductive performance does not seem clear. There was a significant positive correlation between the length of that period and the number of days from calving to first insemination, but the influence of the prolonged acyclic period was not detrimental to the reproductive performance and responsible for only little of the variation. As shown in Table 5 (and Figure 1) it was possible for cows with a long (30-69 days) acyclic period to get pregnant in time to fulfil the criterion for normal reproduction that the number of days open should not exceed 90 days. On the other hand, cows with a short acyclic period could have problems in getting pregnant in time for that (Figure 1).

The relatively small values of R^2 showed that many factors not taken into account must have a strong influence on reproductive performance. These factors include timing of insemination, oestrous behavior, heat detection, time of ovulation, and frequency of early embryonic death.

CONCLUSION

The results presented in this paper are preliminary and based on a small number of animals.

The post partum acyclic period was found to increase with the degree of weight loss early after calving. The acyclic period was not - at least not directly - influenced by milk yield. The intervals from calving to first insemination and from calving to conception (days open) were negatively correlated with the weight loss after calving; the greater the weight loss the longer the intervals and the lower the reproductive performance. The length of the acyclic period post partum was found to be of some influence on the reproductive performance, but most of the cows resumed ovarian cyclicity in time to get pregnant within 90 days post partum. Days of acyclicity post partum is considered to be an insufficient measure of fertility and reproductive performance in dairy cows. The investigation will be continued and repeated.

ACKNOWLEDGEMENTS

We gratefully acknowledge E.Bülow-Skovborg, The Silstrup Research Station, for the opportunity to do the investigations on the herd, and Fritz Christiansen for his very skilful technical assistance. Part of the project was supported by the Federation of Danish AI-Societies.

REFERENCES

Ball, P.J.H. and Pope, G.S. 1976. Measurement of concentrations of progesterone in fat-free cows' milk: Its potential value in studies of reproduction. J. Endocrinology. 69, 40P.

Claus, R., Karg, H., Günzler, O., Müller, S., Rattenberger, E. and Pirchner, F. 1982. Infertility screening post partum using the milkfat progesterone assay.
In: Factors influencing fertility in the post partum cow. Current topics in veterinary medicine and animal science. Vol. 20. Eds. Karg, H. & Schallenberger, E. 288-297.

Henriksen, J., Andersen, O., Nielsen, F. and Pedersen, K.M. 1984. Rapport over projekt "Kvægets Frugtbarhed". National Committee on Danish Cattle Husbandry, DK-8260 Viby. pp 130.

Karg, H. 1981. Physiological impact on fertility in cattle with special emphasis on assessment of the reproductive function by progesterone assay. Livestock Prod. Sci. 8, 233-246.

Oltner, R. and Edqvist, L.-E. 1981. Progesterone in defatted milk: Its relation to insemination and pregnancy in normal cows as compared with cows on problem farms and individual problem animals. Br. Vet. J. 137, 78-87.

Sejrsen, K. and Neimann-Sørensen, A. 1982. Nutritional physiology and feeding of the cow around parturition. In: Factors influencing fertility in the post partum cow. Current topics in veterinary medicine and animal science. Vol. 20. Eds. Karg, H. & Schallenberger, E. 325-357.

Smidt, D. and Farries, E. 1982. The impact of lactational performance on post partum fertility in dairy cows. In: Factors influencing fertility in the post partum cow. Current topics in veterinary medicine and animal science. Vol. 20. Eds. Karg, H. & Schallenberger, E. 358-383.

Webb, R., Lamming, G.E., Haynes, N.B. and Foxcroft, G.R. 1980. Plasma progesterone and gonadotrophin concentration and ovarian activity in post partum dairy cows. J. Reprod. Fert. 59:133-143.

DISCUSSION

Chairman: Karg, H. (FRG)

   Papers and discussion of this session have substantiated the importance of the postpartum situation as the most critical period for fertility in animal husbandry. Professor Vandeplassche stressed in his evaluation of factors interfering with the restoration of reproductive potentials - in recognition of the clinical aspects - the rank of influences whether environmental or of endogenous origin in the mare, cow, ewe, sow, bitch and rabbit. He emphasized in this comparative view not only the marked differences between animal species but also the individuality within these species. Thus general conclusions refering from one species to another have to be avoided. It was elaborated again in the discussion that in cattle the problem of heat detection which is closely related to the management still has top priority for any high level husbandry.

   The endocrinological causes of the peculiar consequences which the farmer has to deal with either in dairy or beef cow operations got explained by the papers of Drs. Terqui and Schallenberger. Dr. Terqui introduced a "new species", the "rustic cow" with weak body condition, which reveals a quite deep and sustained postpartum anestrus. Additionally, he demonstrated convincingly the teaser-bull effect on estrus performance in connection with calf removal but also the sire effect on embryonic death.

   Dr. Schallenberger gave evidence that the dairy cow has developed an alternative strategy to the beef breed animal to maintain a favourable biological rest time of at least 4 - 6 weeks despite the lack of hypothalamo-pituitary quiescene. Ovary and uterus do not resume unimpaired function at an early stage, what postpones fertility, but not necessarily cyclicity. In addition, a new aspect got introduced, the direct interference of the uterus with pituitary function via yet unidentified mechanisms.

   Finally, Dr. Henriksen contributed confirmative evidences of the valuable elucidation of the postpartum period with progesterone profiles and tried to separate out the influences of body condition versus lactation performance on the resumption of fertility.

   I would like to thank the speakers and the participants of the discussion for this vivid session which stressed again that analysis of endocrine events has provided substantial yield of knowledge to face problems of practical importance.

GENERAL CONCLUSIONS

Lamming, E. (U.K.)

The major objectives of the seminar was to summarize and discuss
existing knowledge concerning the problems of seasonal, lactational and
post partum anestrus and these have been achieved. The participants have
reviewed the significance of recent studies designed to investigate individual, breed and species variations in gonadotrophin and steroid hormone
levels associated with these periods of anovulation.

The importance of differences in LH patterns in the breeding and non-
breeding seasons in sheep and horses in relation to ovarian activity has
been emphasized. Information was presented concerning the induction of
the appropriate LH responses and ovulation in sheep by repetitive low
dose GnRH-injections or infusions. These studies indicate a potential
for practical application of GnRH to induce ovulation in farm species.
An important new observation that a GnRH implant can induce ovulation in
the anestrous mare is worthy of special note. In the sheep and cow seasonal changes in prolactin are not considered important in influencing post
partum ovarian activity and seasonal breeding, but in the pig evidence
was presented indicating a possible partial role of prolactin in lactational and post weaning anestrus. Considerable discussion ensued concerning the importance of changes in plasma FSH in relation to follicle
growth and the preovulatory selection of follicles in all species and
evidence presented of biochemical changes occurring in development of
those follicles destined to ovulate compared to those that first develop
and then became atretic. Particular attention was given to the importance
of the LH receptor capacity of granulosa cells and the ability of follicles to undertake conversion of androgens to estrogens. Using evidence
from these studies, parameters suitable to classify the "maturity" of
developing follicles were discussed. Applying these criteria it was reported that some breeds are "mature" follicles in their ovaries during
the non-breeding season and these investigations, which are continuing,
increase our understanding of breed differences in fertility and seasonal
reproductive activity. Important new data on the effects of experimental
light patterns in the mare and in both sexes of the sheep and pig was
presented which increases our understanding of photoperiodic influences
on reproductive activity in these species. As might be expected, new data
on the role of melatonin in relation to seasonal breeding of sheep was
received with considerable interest.

In the session on post partum anestrus the chairman provided a critical
analysis of some previous published information concerning post partum
acyclicity and referred to the contributions of previous EEC-sponsored
symposia in evaluating research investigations within the community. Thus,
the current symposia can be regarded as an update of knowledge. Speakers
in this session emphasized that data from several countries show that a
substantial proportion of herds have "calving to conception" intervals
well in excess of one year. A large number of causal factors are involved
and their effects on post partum physiology and endocrine patterns in
suckled beef and milked dairy cows were analyzed. The speakers re-empha-
sized the critical importance of body condition of nutrition and the
presence of the male on calving to conception intervals. A major feature
of the research reports on post partum physiology in cattle is the uni-

formity between results and views of the different laboratories in the community, a reflection of the value of sponsored interchange of ideas concerning hypotheses and data. Important new data were presented on the possible role of the uterus in the immediate post partum appearance of LH episodes.

The data from all-these sessions collectively provides evidence to indicate why individual animals, breeds and species differ in their response and changes in season, social organization, nutrient supply, lactation and suckling, and assist our development of management techniques to improve the efficiency of livestock production in the community.

LIST OF PARTICIPANTS

BELGIUM

Vandeplassché, M.
Faculty of Veterinary Medicine
State University
Casinoplein 24
9000 Ghent

DENMARK

Henriksen, J.
Institute of Animal Science
Royal Veterinary and Agricultural
University
23 Rolinghedsvej
1958 Copenhagen V

FRANCE

Bour, Barbara
Centre de Recherches Agronomiques
de Tours-Nouzilly
37380 Monnaie

Mauget, R.
CNRS Centre d'Etudes Biologiques
des Animaux Sauvages
Oilliers-en-Bois
79360 Beauvoir-Niort

Terqui, M.
Centre de Recherches Agronomiques
de Tours-Nouzilly
37380 Monnaie

Thimonier, J.
Centre de Recherches Agronomiques
de Tours-Nouzilly
37380 Monnaie

FED. REP. OF GERMANY

Claus, R.
Institut für Tierzüchtung
udn Tierhaltung
Universität Stuttgart-Hohenheim
Garbenstr. 12
7000 Stuttgart 70

Ellendorff, F.
Institut für Tierzucht
und Tierverhalten FAL
Mariensee
3057 Neustadt 1

Dr. F. Elsaesser
Institut für Tierzucht
und Tierverhalten FAL
Mariensee
3057 Neustadt 1

Hoffmann, B.
Ambulatorische und Geburtshilfl.
Veterinärklinik
Universität Geißen
Frankfurter Straße 106
6300 Gießen

Karg, H.
Institut für Physiologie der
Südd. Versuchs- und Forschungs-
anstalt für Milchwirtschaft
Techn. Universität München
8050 Freising-Weihenstephan

Merkt, H.
Institut für Haustierbesamung
und -andrologie
Tierärztliche Hochschule
Bischofsholer Damm 15
3000 Hannover 1

Riemensberger, J. MinR
Bundesministerium für Ernährung,
Landwirtschaft und Forsten
Postfach 140270
5300 Bonn

Schallenberger, E.
Institut für Physiologie der
Südd. Versuchs- und Forschungs-
anstalt für Milchwirtschaft
Techn. Universität München
8050 Freising-Weihenstephan

Smidt, D.
Institut für Tierzucht
und Tierverhalten FAL
Mariensee
3057 Neustadt 1

GREECE

Alifakiotis, Th.
School of Agriculture
University of Thessaloniki
Thessaloniki 54006

IRELAND

Quirke, J.F.
Agricultural Institute
Western Research Centre
Belclare, Tuam,
Co. Galway

Roche, J.F.
Faculty of Veterinary Medicine
University College Dublin
"Kilkieran", Ballycoolin Road
Dublin 11

ITALY

Mattioli, M.
Istituto di Fisiologia Veterinaria
Via Belmeloro 8/2
Bologna

Seren, E.
Istituto Fisiologia Veterinaria
Via Belmeloro 8/2
Bologna

NETHERLANDS

Booman, P.
Research Institute for Animal
Production "Schoonoord"
P.O. Box 501
3700 AM Zeist

UNITED KINGDOM

Allen, W.R.
T.B.A. Equine Fertility Unit
Animal Research Station
307 Huntington Road CB3 OJQ
Cambridge

Foxcroft, G.R.
AFRC Group in Hormones &
Farm Animal Reproduction
University of Nottingham
Sutton Bonington
Loughborough LE12 5RD

Haresign, W.
University of Nottingham
School of Agriculture
Sutton Bonington
Loughborough Le12 5RD

Lamming, E.
Dept. of Physiology
School of Agriculture
University of Nottingham
Sutton Bonington
Loughborough Le12 5RD

McNeilly, A.S.
M.R.C. Reproductive Biology Unit
University of Edinburgh
37 Chalmers Street
Edinburgh EH3 9EW

Wallace, Jacqueline
M.R.C. Reproductive Biology Unit
University of Edinburgh
37 Chalmers Street
Edinburgh EH3 9EW

Webb, R.
A.R.C. Animal Breeding
Research Organization
West Mains Road
Edinburgh EH9 3JQ

C E C

Connell, J.
CEC Direction General de
l'Agriculture, DG VI
200 Rue de la Loi
1049 Brussels
Belgium